Purnell's
Pictorial Encyclopedia of
Nature

Purnell's
Pictorial Encyclopedia of
Nature

Edited by Theodore Rowland~Entwistle

With a foreword by Michael Tweedie

A cheetah and her cubs. As you can see, these animals
are built for speed. They can travel twice as fast as a
top male athlete over short distances. They rely on this
speed for hunting and killing their prey.

Executive Editor:
Campbell L. Goldsmid

Text Consultant:
Michael Tweedie, M.A.
For many years director of the Raffles Museum in Singapore. He is consultant zoologist to *Wildlife* magazine, and has written many books, particularly on insects and dinosaurs. He is a frequent broadcaster on BBC radio.

Photographic Consultant:
Heather Angel, M.Sc., F.R.P.S.
Studied zoology and marine biology before embarking on a career as a photographer which takes her all over the world. She has written many books on natural history and also on the techniques of wildlife photography.

Planned and Compiled by:
Theodore Rowland-Entwistle, B.A., F.R.G.S., F.Z.S.

Editor:
Jean Cooke, B.A.

Additional text by:
Martin Angel, Ph.D.
Neil Ardley, B.Sc.
Jill Bailey, M.Sc.
Alan Jenkins, F.Z.S.
Ann Kramer
David Lambert, M.A.

Designer:
Lindsey Rhodes

Illustrations drawn by:
Stephen Bull
Liz Graham-Yooll
Roger Heaton
Lesley Heron
Valerie Hill
Eric Jewell Associates
Lindsey Rhodes
Cindy Scott
Sam Thompson
Margaret Wells

Designed and produced for Purnell Books by Autumn Publishing Limited, 10 Eastgate Square, Chichester, Sussex

SBN 361 04937 4
Copyright © 1980 Purnell and Sons Limited
Published 1980 by Purnell Books, Berkshire House, Queen Street, Maidenhead, Berkshire
Made and printed in Great Britain by Purnell and Sons Limited, Paulton (Bristol) and London

Contents

FOREWORD

Everyone has heard of the danger that threatens wild nature all over the world, the countryside of the temperate zones and tropical forests alike. Woodlands are felled, swamps are drained, moorland is ploughed and roads and car parks cover more and more of the Earth's surface with concrete. This cannot be avoided if the human race goes on increasing in numbers, and there is no sign that this increase is even slowing down.

Many kinds of animals and plants are probably doomed to extinction, in spite of efforts which are being made to save them by establishing nature reserves – and trying to protect these areas from poaching and so-called development. These efforts can only be effective if the majority of people support them, but people who know little or nothing of the beauty and diversity of living nature will not put themselves out to protect it.

It is not only the larger animals such as rhinoceroses, birds of paradise, otters and badgers that are endangered. Small creatures such as butterflies give just as much pleasure to lovers of nature as the larger beasts, and they are all in retreat, the great bird wing swallowtails of the Papuan rain forests and the little blue butterflies of the European grasslands alike.

This book gives an overall picture of the natural world, both animal and plant life. If you read it carefully and look closely at the photographs which illustrate it you will, we hope, be persuaded that wild nature is beautiful and marvellous and well worth saving. If you already hold this opinion the book will give you real enjoyment.

Michael Tweedie

These are Coke's hartebeeste. Several species of hartebeeste are in danger of extinction.

THE MARVEL OF LIFE

Living things are the subject of this book. As far as we know at the moment, Earth is the only planet in the Solar System that has life on it. We know there is no life on the Moon, and conditions appear to be impossible for life on most of the other planets. Mars is the only other planet where life might be possible, and so far no one has been able to detect it there. Scientists think that conditions have to be exactly right for life to exist, and Earth is the only one of the nine planets circling the Sun which has those conditions. The Sun is a star, and there are millions of other stars in the universe. We suspect that some at least of these stars must have planets, and that some of them must have conditions identical to those on Earth, but for us at the moment, the marvel of life is confined to our own planet.

It is much easier to identify living things than to say what life is. Animals, plants, bacteria too small for the eye to see without aid – all these are living. We know they are living because they grow, they need some kind of food, and they reproduce themselves so that there is always a new supply of living things. Life itself – what it is and why it is – remains a mystery.

However, scientists have found out a great deal about the mechanics of life, and even how it probably began on Earth. This, the first section of the *Pictorial Encyclopedia of Nature*, looks at life and nature and explains how best to understand them.

Life is just beginning for this chicken. Although it has broken the protective shell its feathers are still wet. Soon they will dry and it will be recognisable as a yellow, fluffy chick.

How Life Began

Everything in the Universe is made up of a number of basic chemicals, called elements. Scientists have found 92 of these elements in Nature, and have been able to make more than a dozen others – most of which change very rapidly and have only a short life. All these chemicals exist mainly in mixtures known as compounds. For example, the elements hydrogen and oxygen combine to form a world-wide compound – water. There are two kinds of compounds, organic and inorganic. Organic compounds are found only in living things or the products of living things. They always contain the element carbon, so it is clear that carbon is an essential element in life. Inorganic compounds can be found both in living and non-living things – water is an example. Some inorganic compounds also contain carbon.

These simple facts about chemistry are essential to understanding how life is organised and how it probably began. The Earth is thought to be about 4,600 million years old. We know that life began at least 3,000 million years ago. Figures like these are very difficult to grasp: they are so big. If we imagine that the Earth was made one year ago, then life began about 34 weeks ago, mammals first appeared less than two weeks ago, and Man has been in existence only a few hours.

The formation of amino-acids marked the beginning of life on Earth.

One-celled creatures like these were the first forms of animal life on Earth.

Written history has covered only a matter of a few minutes.

Scientists think that the Earth was originally a whirling mass of gases, and very hot. As it cooled down it gradually assumed its present shape, with a hard outer crust of rocks, and hotter rocks and metals deep inside. All this time enormous chemical changes were going on, as the elements of which the Earth is made combined and recombined to form different compounds. Among these compounds was water: in other words, the great oceans were formed.

Gradually chemical changes in the oceans built up compounds which could be used as food by living organisms. At some stage a group of chemical compounds called amino-acids evolved. Amino-acids contain the elements carbon, hydrogen, oxygen and nitrogen. Long chains of amino-acids linked together to form proteins, more complex compounds which have been called the building blocks of life.

With so much potential food around in the sea, eventually the amino-acid compounds started absorbing the food and performing the chemical and physical change we know as growing. In other words, life had begun. Although we do not know for certain that life did begin exactly like this, we do know that something of the sort must have occurred.

The starting of life in this way is known as spontaneous generation. At one time scien-

Trilobites

Brachiopods

Eusthenopteron

Diplocaulis

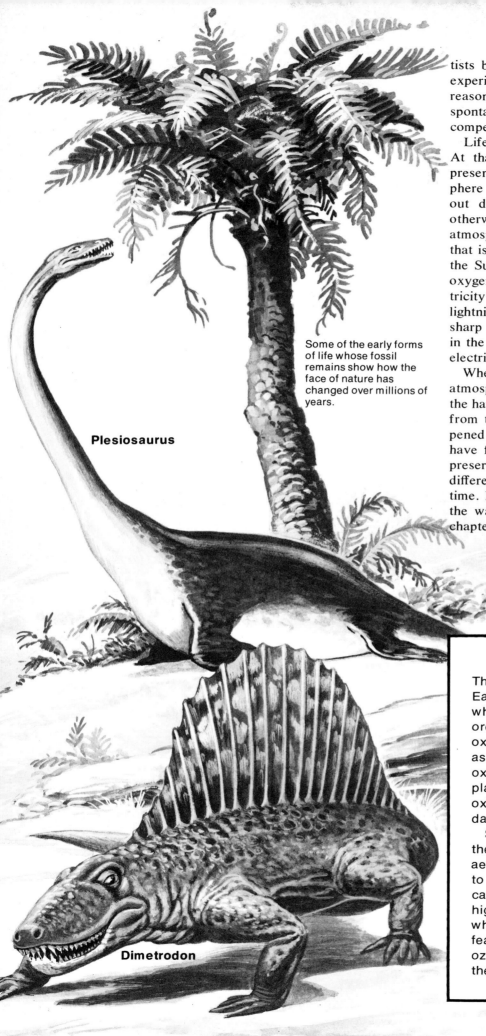

Some of the early forms of life whose fossil remains show how the face of nature has changed over millions of years.

Plesiosaurus

Dimetrodon

tists believed that it was still going on, but experiments have shown that it is not. The reason is probably that the conditions for spontaneous generation just cannot exist in competition with the life that already exists.

Life stayed in the sea for a very long time. At that period the Earth did not have its present blanket of air around it. This atmosphere has many uses. One of them is to filter out dangerous rays from the Sun, which otherwise would shrivel everything up. The atmosphere contains many gases, but the one that is the most important as a shield against the Sun's rays is ozone. Ozone is a form of oxygen, which is made by the effect of electricity on the oxygen in the air. Every flash of lightning helps to make ozone. This gas has a sharp smell, and you can sometimes detect it in the air after a thunderstorm, or even near electric motors.

When enough ozone had formed in the atmosphere to make a good protection against the harmful rays of the Sun, life could emerge from the sea on to land. We think this happened about 420 million years ago, because we have found fossils (remains of living things preserved as rock) which show that all sorts of different forms of life began to evolve at that time. For more information about fossils and the way living things have changed see the chapter on *Prehistoric Life*, on pages 162-173.

Producing oxygen

The most important gas in the Earth's atmosphere is oxygen, which all animals must breathe in order to survive. Living plants put oxygen into the air. They use water as part of their food, and give off oxygen in the process. In this way plants have helped to put enough oxygen into the air to make present-day life possible.

Some scientists are worried that the chemicals used in modern aerosol spray cans may be a threat to life. Some of these chemicals, called fluorocarbons or freons, rise high into the Earth's atmosphere when they are released. There, it is feared, they may break down the ozone layer which protects us from the Sun's rays.

Plants, Animals - and Others

The world of living things is divided into two great groups: the Animal Kingdom and the Plant Kingdom. As a rule it is easy to tell a plant from an animal: plants are green and stay in the same place: animals are not green and can move about. These are very simple descriptions of plants and animals, and scientists would tell you that they are not invariably correct: for example, not all plants are green, and some animals are. Some animals do not move about. However, there are very important differences between plants and animals, and there are amazing similarities too.

One of the most important differences cannot be seen: plants can make their own food, using water and minerals which they abstract from the soil, the gas carbon dioxide which they take from the air, and sunlight as a source of energy to do the work. This process is called photosynthesis, and you can read more about it on pages 112-113. Animals cannot make their own food, so they have to eat plants to get nourishment. Some animals get their plant-food secondhand, as it were, by eating other animals which have fed on plants.

Another important difference which you cannot see except with a microscope is in the structure of the cells from which plants and animals are made. Just as proteins have been called the building blocks of life itself, so cells are the building blocks of living things. You can read about cells on pages 14-15.

The ways in which plants and animals grow differ. Most animals grow until they reach their full size and then stop. A full-sized animal is known as an adult. Many plants continue to grow long after they have reached maturity, some of them for hundreds of years. A tree, for example, puts on a new layer of tissue all round every year. You can see this growth when you look at the stump of a tree that has been cut down. Each year's extra material shows up as a ring, and by counting the rings you can tell how old the tree was. Some plants die down and then grow up again, but animals never do this.

Both animals and plants respond to their surroundings, but animals are much more sensitive than plants. Plants are affected by only the simplest things, such as heat, moisture and light. You can see that a plant responds to light if you stand one on a window ledge. Most of its leaves and flowers turn to face the light, and if you turn the plant round so that the flowers face away from the window, the plant will gradually turn them back on their stems.

Animals have a much greater awareness of what is around them. The lower forms of

Far right: Plants respond to stimuli in very simple ways. These cress seedlings have turned their leaves to face the source of light.

Right: A wild cat responds actively to stimuli from its environment: sight, smell, and hearing are all brought into play.

animal life make what are merely instinctive responses to stimuli – that is, anything that stimulates the senses. The higher animals have well-developed senses of touch, smell, hearing, sight and taste. Although they all make many instinctive reactions – just as you snatch your hand away if it touches something very hot – a great many of these reactions require thought and choice. There is more about instinct on pages 90-91.

Right: An otter moves around to catch its food – a fish – while a plant makes its own food.

Below: Animals can move about, plants cannot. A group of elephants rests under a pair of trees. The elephants can move off whenever they wish – but the trees stay put.

A third kingdom?

Scientists cannot be sure whether some of the smallest living things are true plants or true animals. So there is a growing move to have a third kingdom, the Protista, into which these borderline organisms can be fitted.

The Protist Kingdom would contain bacteria, which are a bit like algae and a bit like the simplest one-celled animals, the protozoans. It would also contain the blue-green algae, the protozoans, the other algae, and fungi. Viruses, which are also difficult to classify, are not included because they do not appear to be capable of independent life outside the cells of some other organism. Viruses produce many kinds of infectious diseases.

Bacteria

Bacteria are tiny organisms which can be seen only with a microscope. There are more bacteria than any other living things, and they can be found everywhere. Many live in the digestive organs of animals, where they help the process of digestion.

Some bacteria break down the dead remains of plants and animals into simple chemicals that can be used again. Others extract nitrogen gas from the air and put it into water and the soil, where it is vital for plant growth. Bacteria help to convert sewage into safe substances. They also play a part in fermentation, the process used to make certain foodstuffs such as cheese.

Some bacteria, however, are harmful to us. They can make food go bad and so cause food poisoning. Others cause some of the major diseases, such as pneumonia, tetanus and typhoid fever.

Units of Life

The word cell originally meant a small room, and we still use it in that sense when we speak of a monk's cell or a prison cell. It is a very good word to describe the biological cell, the basic unit out of which all living things are constructed. For a cell is a complete miniature organism in itself, surrounded by a soft, flexible membrane. Plant cells have a stiff cellulose wall as well as the flexible membrane. The cellulose helps to support the plant.

A few animals and plants consist of one cell only – you can read about them on pages 88-89 and 136-137. Larger animals are built up from many cells of differing types, each with its own particular job to do. A man's body contains approximately one million million cells, each so small that you can see it only under a microscope.

The inside of a cell is filled with a jelly-like substance called protoplasm. This protoplasm is very complicated. It consists of two main divisions: the nucleus and the cytoplasm. The nucleus is comparatively small and enclosed by its own membrane. The nucleus is the cell's control centre: it directs the cell's activities and it contains all the information needed to

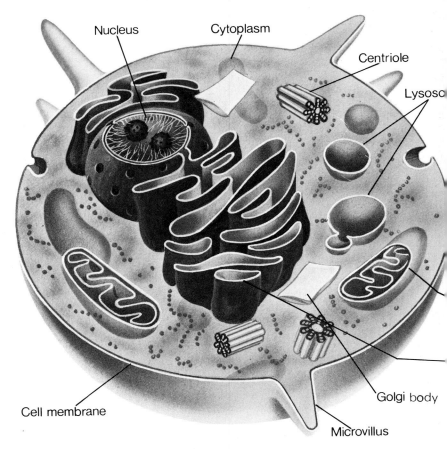

Animal Cell

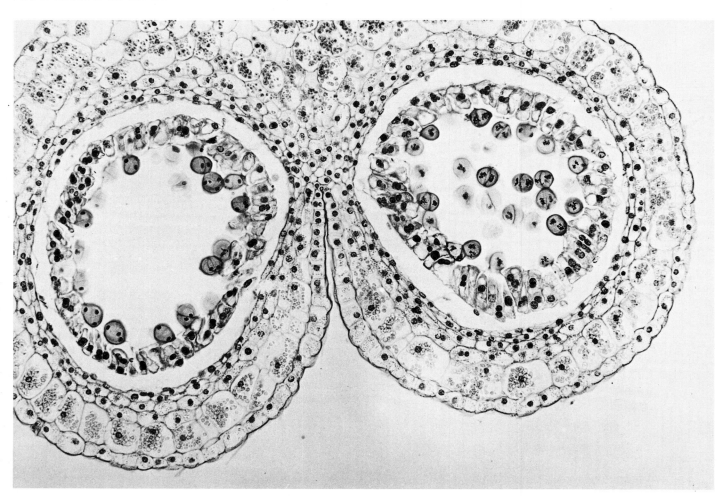

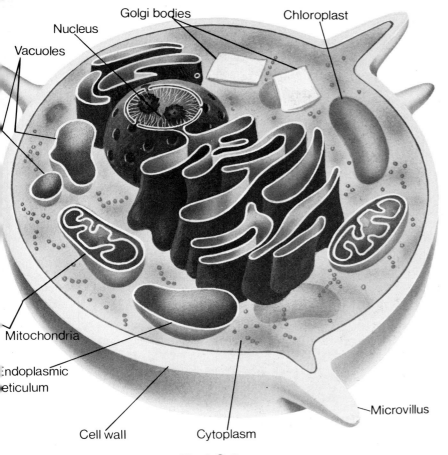

Golgi bodies
Chloroplast
Nucleus
Vacuoles
Mitochondria
Endoplasmic reticulum
Cell wall
Cytoplasm
Microvillus

Plant Cell

keep the contents of the cell at pressure, and so assist in making it rigid – rather as air in an inner tube makes a bicycle tyre rigid.

Nearly all the cells in your body have a limited life. They die and are replaced by fresh cells. Some live only a few days, others for several months. Millions of cells die every day: skin cells flake off, while internal cells are passed out of the body with other waste matter. Nerve cells live the longest, but if they die they are not replaced. This is why damage to the nerve cells of the brain is often so disastrous, while damage to other parts of the body can be repaired. However, the body does its best to compensate for damage even to the brain, and often other cells take over the work of those injured.

There are two ways in which cells reproduce. Most cells do it by a process known as mitosis. In mitosis a cell splits, producing two identical 'daughter cells'. These cells in turn grow to their full size and can then divide. When a plant or animal is growing, more cells are formed than die. When the animal is fully grown, just enough cells are produced to maintain the body. Most of the work involved in cell division goes on inside the nucleus, and is described on pages 16-17, where you can also read about the other form of cell division, which is called meiosis.

Above: These diagrams show in simplified form what the inside of (left) an animal cell and (right) a plant cell is like.

Right: Stages in cell division by mitosis:
1 Cell before division;
2 Chromosomes and the centrioles duplicate themselves and pair up.
3 The nuclear membrane breaks, and the chromosomes move into the centre of the cell.
4 Each chromosome pair splits in two.
5 A new nuclear membrane forms around each group of chromosomes, and two new cells are formed.

Left: A microphotograph showing cells of a lily dividing. These are sex cells, and they are at the second stage of division by meiosis.

make sure the cell does exactly its own job, and nothing else. Details of the marvellous work that goes on inside the nucleus are given on pages 16-17.

The rest of the cell is filled with cytoplasm, which is mostly water with tiny structures called organelles in it. The power plant of the cell is provided by hundreds of mitochondria, which are rod-shaped organelles. Centrioles look like bundles of rods, and they play a part in cell division. They are not found in the cells of most plants. Lysosomes are spherical and are found only in animal cells. They help to destroy the cell when it dies. A network of membranes forming channels is found in most cells, and scientists think it acts as a sort of communications network. It is called the endoplasmic reticulum. Finally there are the Golgi bodies, flat sac-like structures whose work is still not fully understood. In plants it is thought they help to produce the cell wall material.

In addition to these structures plant cells contain organelles called chloroplasts. These organelles contain the green substance chlorophyll, which makes plants green. Other, similar organelles known as chromoplasts hold those pigments, such as red, orange and yellow, which give flowers their colours. Vacuoles are membranous sacs filled with a watery fluid, found in plants. They help to

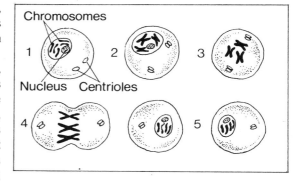

Chromosomes
1 2 3
Nucleus   Centrioles
4 5

Cellulose

The cellulose which makes up the cell walls of most plants (not fungi, nor some of the algae) is a tough substance which plants put together from carbon, hydrogen and oxygen. About one-third of all plant matter is made up of cellulose. Animals such as cows and horses can digest cellulose with the aid of bacteria which live in their digestive systems. Human beings cannot absorb it, but it provides useful 'roughage' which helps the digestive system in its work.

The Code of Life

As explained on pages 14-15, the nucleus is the control-centre of the cell. It regulates the moment-by-moment activities of the cell and also the way in which it reproduces itself. In the nucleus is all the information needed to make sure that the cell, and any daughter cells formed from it, is of a particular kind. Cells are highly specialised: not only is a cell from an animal different from that of a plant, but cells of each species are different from those of every other species – and cells within an organism differ, too. For example, muscle cells are very different from brain and liver cells.

The chemical substance that acts as the controlling agent is called deoxyribonucleic acid, or DNA for short. DNA is one of the two main constituents of the cell nucleus. The other substance is protein, and each nucleus contains many kinds of proteins. DNA and the proteins together make up a substance called chromatin. Besides chromatin, the nucleus contains little round bodies called nucleoli. These bodies also contain proteins, and another chemical substance called ribonucleic acid – RNA for short.

To understand how a cell produces two identical cells, it is necessary to look at the mechanics of ordinary cell division (mitosis). The division begins with the chromatin forming into a number of rod-like shapes called chromosomes. The number of chromosomes varies according to the kind of organism: for example, a human body cell contains 23 pairs, while a potato cell has 24. The vinegar fly has just four pairs, and one species of thread worm only two. The cells of one kind of fern contain 250 pairs of chromosomes.

The chromosomes and the centrioles, organs like bundles of rods (see pages 14-15), then duplicate themselves, and move to opposite ends of the cell, out of the nucleus. The cell then splits in two, with each half having the same original number of chromosomes. The new cells are absolutely identical to the original cell.

The chromosomes carry all the essential information for the function of the cell. The DNA consists of chains of organic acids arranged in a special order. This order is a code carrying the information. The cell is programmed like a computer or a calculator, except that the cell's programme is recorded chemically and the calculator has an electronic code.

There is a second method of cell division, called meiosis. Certain cells of an animal or plant, known as sex cells, divide by meiosis. In this the first stage is similar to mitosis: the

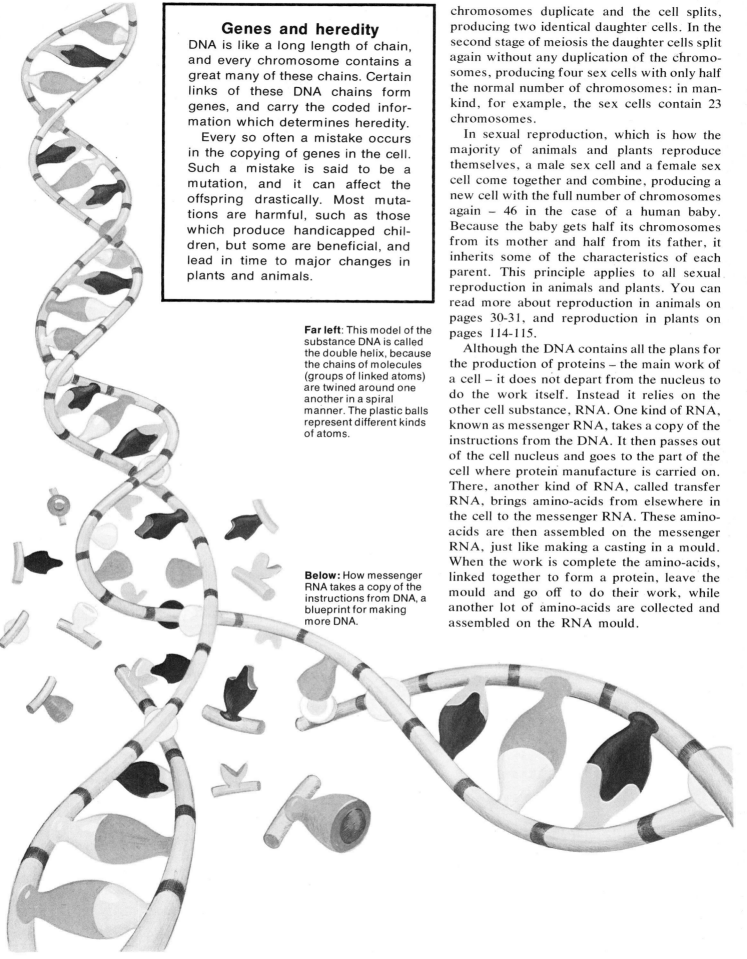

Genes and heredity

DNA is like a long length of chain, and every chromosome contains a great many of these chains. Certain links of these DNA chains form genes, and carry the coded information which determines heredity.

Every so often a mistake occurs in the copying of genes in the cell. Such a mistake is said to be a mutation, and it can affect the offspring drastically. Most mutations are harmful, such as those which produce handicapped children, but some are beneficial, and lead in time to major changes in plants and animals.

Far left: This model of the substance DNA is called the double helix, because the chains of molecules (groups of linked atoms) are twined around one another in a spiral manner. The plastic balls represent different kinds of atoms.

Below: How messenger RNA takes a copy of the instructions from DNA, a blueprint for making more DNA.

chromosomes duplicate and the cell splits, producing two identical daughter cells. In the second stage of meiosis the daughter cells split again without any duplication of the chromosomes, producing four sex cells with only half the normal number of chromosomes: in mankind, for example, the sex cells contain 23 chromosomes.

In sexual reproduction, which is how the majority of animals and plants reproduce themselves, a male sex cell and a female sex cell come together and combine, producing a new cell with the full number of chromosomes again – 46 in the case of a human baby. Because the baby gets half its chromosomes from its mother and half from its father, it inherits some of the characteristics of each parent. This principle applies to all sexual reproduction in animals and plants. You can read more about reproduction in animals on pages 30-31, and reproduction in plants on pages 114-115.

Although the DNA contains all the plans for the production of proteins – the main work of a cell – it does not depart from the nucleus to do the work itself. Instead it relies on the other cell substance, RNA. One kind of RNA, known as messenger RNA, takes a copy of the instructions from the DNA. It then passes out of the cell nucleus and goes to the part of the cell where protein manufacture is carried on. There, another kind of RNA, called transfer RNA, brings amino-acids from elsewhere in the cell to the messenger RNA. These amino-acids are then assembled on the messenger RNA, just like making a casting in a mould. When the work is complete the amino-acids, linked together to form a protein, leave the mould and go off to do their work, while another lot of amino-acids are collected and assembled on the RNA mould.

Studying Natural History

The study of natural history, or biology to give it its more formal name, has been going on since the days of ancient Greece. The first great biologist was the philosopher Aristotle (384-322 BC); the two terms philosopher and scientist shared the same meaning until comparatively recently. Many Greeks and Romans studied nature; for example, the Roman scholar Pliny the Elder (AD 23-79) wrote an encyclopedia of natural history.

During the Middle Ages fewer people studied biology, but there was a revival of interest from the 1400s onwards. In the mid-1700s, as you can read on pages 22-23, the Swedish botanist Carolus Linnaeus laid down

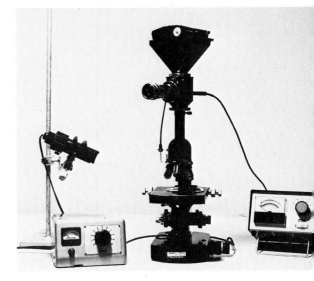

Right: The apparatus for taking micro-photographs. The microscope has a camera on top holding the film. On the left is a transformer supplying current to the lamp above it, and there is a light-meter on the right.

Below: Gathering specimens of freshwater life from a pond, using a very fine mesh net.

the principle of classifying plants and animals, and in 1859 the British naturalist Charles Darwin described his theory of evolution, which is explained more fully on pages 170-171.

In the early days of science people were able to study all aspects of biology because there was only a limited amount of knowledge available to them. Today each subject is so complex that biologists have to specialise. Those interested in plants study botany; those who prefer animal life follow zoology. Even within one discipline scientists now specialise: for example, entomologists study insects, marine biologists examine life in the sea, and microbiologists study the smallest forms of life, such as bacteria.

Biologists today have many sophisticated pieces of apparatus to help them in their work. Their chief aid is the microscope, which enables them to examine not only very small organisms but also the details of much larger ones. The more elaborate electron microscope produces a picture that is magnified up to 800,000 times life size. A centrifuge, which whirls mixtures round at high speed, is used to separate substances of different weights, such as large and small bacteria.

Work with elaborate machines can be done only in a laboratory, but a great deal of a biologist's work is done in the field, observing objects in their natural surroundings. By travelling around biologists can observe how plants grow in different conditions, how animals live, find their food and rear their young, and how changes affect them.

Many biologists travel to collect specimens, either to bring back dead for study in the laboratory, or alive to keep in zoos and botanical gardens. Others travel to photograph living things: for example, the naturalists who took the photographs in *Purnell's Pictorial Encyclopedia of Nature* travel all

Above: A hide, where a photographer or observer can shelter and move about without disturbing the animals being studied.

Left: A group of entomologists collecting insects which have been attracted to a canvas screen.

over the world to get pictures of plants and animals. One thing nature photographers, and indeed all biologists, need is patience: if you want to see animals in their natural surroundings you have to wait for them to appear, and they often seem to be particularly elusive just when somebody wants to watch them.

That is one reason why zoos and nature

Enthusiasts looking for specimens of marine plant and animal life on a rocky shore.

reserves have an important part to play in the study of natural history. A zoologist can study an animal in captivity much more closely than in the wild; but he may get a false impression of its behaviour, just as he would if he observed a farmer in a city street. However, modern zoos are being designed to reproduce as closely as possible, if on a small scale, the conditions in which the various animals live. For example, the Bronx Zoo in New York City has a section called the World of Darkness, where specially dim lighting recreates the night-time atmosphere in which many animals are active. Other zoos, such as the Ibadan Zoo in Nigeria, use dry moats and water barriers rather than cages to keep the exhibits from getting out, so that visitors can see them more clearly. Whipsnade Zoo, an out-of-town branch of the London Zoo, was one of the first of these 'natural' zoos when it was opened in 1931. London Zoo itself has the largest number of different species of animals of any zoo.

Botanical gardens perform much the same function for plants as zoos do for animals. Botanists in temperate climates have to keep tropical plants in hothouses, where both the temperature and the humidity – the amount of moisture in the air – are controlled to reproduce tropical conditions as closely as possible. Nature reserves are areas in a country set aside for the wild life – plants and animals – of that region. Studying animals there is not so easy as in zoos, but scientists can see them in their natural environment. Nature reserves go under a variety of names – national parks, game reserves, even safari parks – and there are more than 200 major reserves world-wide, besides many smaller ones. Some national parks attract so many visitors that even there wild plants and animals are imperilled. See also pages 180-181.

Tree gardens

One specialised kind of botanical garden is the arboretum, which is a plantation devoted entirely to trees. In a well laid out arboretum specimen trees collected from all parts of the world are grouped attractively in a park-land setting.

Because trees are among the most hardy of plants it is possible to grow together trees from a wide variety of climates, provided that the soil is suitable. Many arboretums include a selection of shrubs among the trees, to give a more natural effect.

Natural History for All

Natural history is a subject that anyone can study anywhere. Even if you live in one of the concrete jungles which are our modern cities, you can still find plenty of plants and animals to observe. Although access to a good laboratory and time and money to travel are an immense help to biologists, a lot can be learned by using simple equipment. Natural history students living in a city have one great advantage over country-dwellers: access to libraries, where they can find out more about their subject.

Many people begin their natural history studies by bird-watching, because birds find a living almost everywhere. Binoculars are helpful for observing birds at a distance, but not essential. It is just as rewarding to attract birds to you by providing them with food, nesting materials and protection from cats, so that you can take a close look at their way of life. Then you can learn to identify the birds – and when you try to tell them apart remember that the males and females often have different colouring. Listen to the calls of the birds. There are plenty of recordings made to help people identify bird song. If you live in a quiet

Dewdrops in the early morning help to make this spider's web show up its intricate patterns. A natural historian has to make observations at all times of the day.

Below: Binoculars are a great help in studying animals. Because the observer can watch from a far greater distance, he or she can observe the animals closely without disturbing them. It is best to have a fairly lightweight pair of binoculars because it is much easier to hold them steady.

Below: A wire cage of nuts hung by a bird table or near a window attracts birds such as tits which can hang upside down to feed on them. In this way you can lure the birds to you so that you can observe and study them more closely.

area you could record birds, as well as animals and insects, yourself. Obviously it is easier to learn if there is somebody to point out the various species, but you can find out a great deal on your own by looking and listening.

Even more accessible than birds are insects, such as flies and beetles, and the many kinds of spiders. Spiders do a lot of good to mankind because of the enormous quantity of harmful insects that they eat, but for some reason quite a number of people are frightened of them. One distinguished zoologist began his career as a boy by identifying spiders he found near his school, discovering many species which had not previously been noted in the area. His example shows an important point in studying natural history: keep notes of what you see and find. Often important discoveries are made by comparing what is seen in one place and at one particular time with later observation.

When you come across something interesting you will want to look at it in detail. It may be difficult to get hold of a microscope, but a small hand lens carried in a pocket means you can examine a specimen closely as soon as you have discovered it. Write down or draw any special feature that you consider worth noting, then let the creature go again to carry on with its normal way of life.

Whenever you get a chance to visit a zoo or a nature reserve, do so. It may not be as exciting as going on safari in the wild, but seeing a living animal that is well looked after adds to understanding the articles you can read about it in books and magazines. Most zoos have information available about their exhibits, and some have guided tours for people who are really interested.

If you prefer to study plants, you have even more opportunities because plants grow in the most unlikely places. You can raise plants for yourself, either in a garden or indoors in pots, and when you are out you can keep your eyes open for interesting specimens. Photography is an important way to keep a record of your finds, and at least plants do not try to run away while you take their pictures. You may be able to visit botanical gardens to see collections of exotic plants growing in special conditions such as hot, humid greenhouses for tropical plants, and dry, stony areas for cacti.

The next best thing to seeing living plants and animals is to make use of broadcasts. Television and radio often have unusual programmes on natural history subjects, and television in particular gives viewers a chance to see living things in their natural habitats, places where they would not normally be able to visit. Programmes which follow a living thing all through its life cycle show at one time the stages of its development.

Pets such as a domestic cat make fascinating subjects for study, as well as good company. You can learn a lot about the cat family in general by watching your own pets and making notes on their appearance, their habits and their likes and dislikes.

Keeping pets

Many people like to keep animals as pets, but an animal in the home can be much more: it can be a subject for study. If you keep a careful watch on its activities, you can learn an amazing amount about its normal behaviour and its reactions to its surroundings.

The most popular pets are dogs and cats, but there are many other animals that can be kept in the home, and some of them offer many fascinating possibilities for study. For example, a tank of tropical fish provides opportunities for studying the fishes' eating habits, how they swim, their reactions to one another and their breeding habits.

However, whatever animal or animals you decide to keep, remember that owning a pet is a responsibility. You must look after it properly, and see that it has the right food and living quarters. A dog needs regular exercise; a cat must have the chance to prowl; a small animal such as a hamster or a canary must have its cage kept clean all the time.

What's in a Name?

If you were to speak about a robin to an international group of bird-watchers, they would not all have the same mental picture. At least ten different birds are called robins, including the robins of Europe and North America, the Pekin robin of the Himalayan foothills of India, the black robin and pied robin of Sri Lanka, and five flycatchers in Australia. On the other hand, the same bird can have several different names, depending where you are: the thrush, in English, is *die Drossel* in German, *la grive* in French and *il tordo* in Italian. The same confusion can be found in plant names, too.

To avoid this confusion biologists the world over use scientific names to describe animals and plants. The names are written in a form of Latin, which can be readily understood in any country. Scientific names are a way of classifying plants and animals, so that those which are of similar type are grouped together. Classification makes it easier to study botany and zoology. The science of classification, as applied to animals and plants, is called taxonomy, from a Greek word meaning to arrange.

The modern system of classification was begun in the late 1600s by an English naturalist, John Ray, who is often called the 'father of natural history'. It was continued and fully developed by a Swedish naturalist, Karl von Linné, better known by the Latin version of his name as Carolus Linnaeus. It has been extended by later biologists, and has changed as people have learned more about the animals and plants they are classifying.

Above: Carolus Linnaeus (1707-1778), drastically altered the methods of naming and classifying living things, and founded the system in use today.

Right: *Pan satyrus*, more commonly known as the chimpanzee, is one of the most intelligent and most popular of the ape family.

Below right: The common oak, *Quercus robur*, is a deciduous tree which flourishes in temperate climates.

The basis of all classification is the species, that is the individual type of living thing. Members of the same species are generally very much alike, though there may be subspecies. As a rule, members of different species do not mix together to breed. A daisy is always a daisy, and a robin is always a robin. Species that are closely related belong to a genus, and several genera (the plural form) constitute a family.

Altogether there are seven basic taxons – that is, taxonomic groups, or ranks. The biggest groups are the kingdoms, of which there are two, Animalia and Plantae (animals and plants). Some scientists think there should be a third kingdom, the Protista, for some living things that are not true plants or animals, but a bit of both (see pages 12-13). Each kingdom is divided into major groups called phyla (singular phylum) for animals, and divisions for plants. Each phylum and division contains classes; each class is divided into orders, each order into families, and so to genera and species.

In addition to these main groupings, scientists often use other ranks to help them sort out some of the more complicated animal and plant groups, such as super-class, sub-order, sub-family and so on.

By convention, all the ranks down to family are printed in roman type with an initial capital letter; for example the order Carnivora, flesh-eating animals, and the plant division Bryophyta, which includes mosses and liverworts. However, in referring to an individual kind of animal or plant scientists generally use only the genus and species, and again by convention these are always printed in *italic* type, with a capital letter for the genus but not

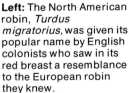

Left: The North American robin, *Turdus migratorius*, was given its popular name by English colonists who saw in its red breast a resemblance to the European robin they knew.

Right: The European robin, *Erithacus rubecula*, is a much smaller bird than the North American robin. Both species of birds belong to the thrush family, Turdidae.

for the species. For example, the polar bear is genus *Thalarctos*, species *maritimus*. Very often the genus and species name describe the character of the animal or plant: for example, *Thalarctos maritimus* means sea-going animal of the Arctic Ocean. Sometimes the species name commemorates the person who first discovered it.

Biologists use a form of shorthand: if they are referring to more than one species of the genus *Felis*, for instance, having named the first one in full *Felis domestica* (the domestic cat) they refer to the others as *F. sylvestris* (the European wild cat) or *F. viverrina* (the fishing cat). If they do not want to be precise as to species they may just say *Felis* sp.

In the ordinary way you do not need to use the scientific name for the more common animals and plants, but for a great many, particularly among the invertebrates, there is no common name; the Latin name is the only one. For this reason you will find a number of Latin names in this book.

THE ANIMAL KINGDOM

Animals are the living creatures most of us understand best, because we ourselves are part of the Animal Kingdom. Few people except zoologists, the people whose business is to study animals, realise fully the incredible variety of animal types, each one perfectly adapted for the life it leads. Animals live in almost every part of the world, a few species even making a chilly home on the edge of Antarctica. That bleak and terrible place is otherwise the only region of the world where animals cannot survive naturally. Only Man has been able to penetrate the continent's interior.

The most familiar animals are the ones which, like ourselves, have backbones – the vertebrates. Possession of a backbone makes the body stronger, and enables its owner to grow large. Yet the vast majority of animals are invertebrates, having no backbones. They are small, and they can make their homes in places and conditions where no large animal could survive. The different kinds of vertebrates are numbered in tens of thousands, but nobody knows how many varieties of invertebrates there are: the insects alone have more than one million species.

In the next 82 pages we look at all aspects of the Animal Kingdom, from the mammals, the group to which we belong, working back down the animal family tree to the simplest creatures. We also take a look at some aspects of animal ways of life, and we begin with a quick glance at animal types and bodies.

A group of African elephants refresh themselves with water. Water is essential for all forms of animals, and they often travel long distances to reach it.

Animal Groups

There are 23 phyla, or major divisions, of the Animal Kingdom. These phyla differ considerably from one another in the structure of the bodies, yet strangely nearly all the animals that are most familiar to us come from about half-a-dozen of these phyla. Most of the rest are small creatures living in or near the sea.

Chordata are the most important because they include all the vertebrates – those animals with backbones, including Man.

Arthropoda are animals with a segmented external skeleton and six or more jointed legs. They are the most numerous of all animals, and include spiders, crustaceans, insects, centipedes and millipedes.

Echiuroidea are soft, fleshy sea animals, which burrow in sand and mud.

Priapuloidea are sausage-shaped sea animals, also living in mud.

Sipunculoidea are seashore animals with slender bodies.

Annelida are the segmented worms, of which the earthworms are the best known.

Mollusca are the familiar snails and slugs, and other animals with shells such as clams.

Chaetognatha, the arrowworms, are small, transparent sea creatures.

Echinodermata are spiny-skinned animals mostly shaped like five-pointed stars; they include sea urchins and starfishes.

Brachiopoda are called lamp shells from their shape. They cling to rocks in the sea.

Pogonophora are long, thread-like worms living deep in the ocean.

Phoronidea are sea animals living in tubes which they build in mud.

Acanthocephala are tiny parasitic worms which live in the guts of larger animals.

Aschelminthes are small, worm-like animals including wheel worms and roundworms.

Polyzoa (moss animals) cannot move around, but live in colonies on rocks.

Entoprocta are like miniature flowers, less than 1 millimetre long. Most live in the sea.

Nemertinea are ribbon worms, flat and very long – up to 27 metres (90 feet).

Platyhelminthes are flatworms, most of which are parasites. They include the flukes and tapeworms which can live inside people.

Ctenophora, also called comb jellies or sea walnuts, live in the sea.

Coelenterata are jelly-like animals, including jellyfish, sea anemones and the corals.

Mesozoa are tiny parasites which live in octopuses and squids.

Porifera, also called Parazoa, are the sponges.

Protozoa are the simplest animals of all, each consisting of one cell.

Land snail

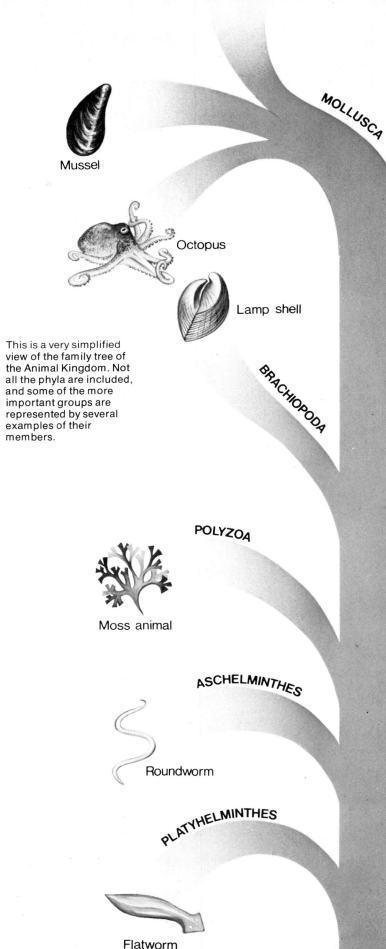

Mussel

Octopus

Lamp shell

This is a very simplified view of the family tree of the Animal Kingdom. Not all the phyla are included, and some of the more important groups are represented by several examples of their members.

MOLLUSCA

BRACHIOPODA

POLYZOA

Moss animal

ASCHELMINTHES

Roundworm

PLATYHELMINTHES

Flatworm

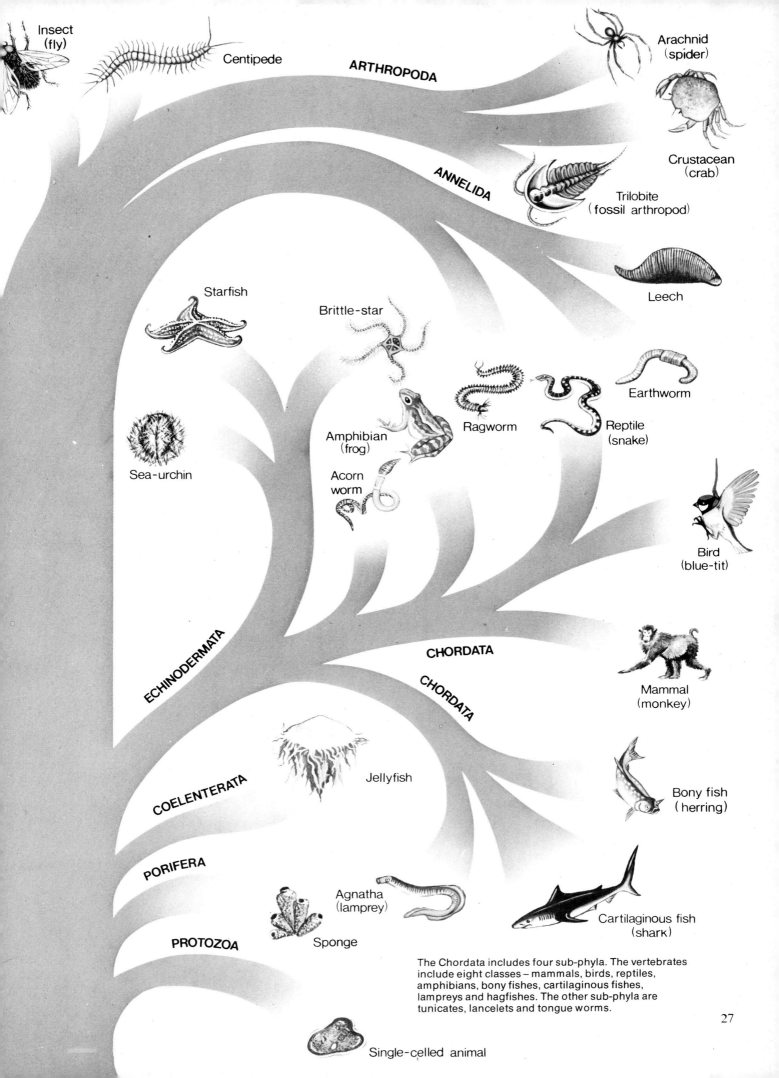

Insect (fly)

Centipede

ARTHROPODA

Arachnid (spider)

Crustacean (crab)

ANNELIDA

Trilobite (fossil arthropod)

Leech

Starfish

Brittle-star

Earthworm

Ragworm

Reptile (snake)

Amphibian (frog)

Sea-urchin

Acorn worm

Bird (blue-tit)

ECHINODERMATA

CHORDATA

CHORDATA

Mammal (monkey)

Jellyfish

COELENTERATA

Bony fish (herring)

PORIFERA

Agnatha (lamprey)

Cartilaginous fish (shark)

Sponge

PROTOZOA

The Chordata includes four sub-phyla. The vertebrates include eight classes – mammals, birds, reptiles, amphibians, bony fishes, cartilaginous fishes, lampreys and hagfishes. The other sub-phyla are tunicates, lancelets and tongue worms.

27

Single-celled animal

Animal Bodies

Animal bodies vary enormously – in shape, in size, in appearance generally – but all animals have certain needs in common. They must take in oxygen, which they need in order to obtain energy. They must take in food, and to obtain their food nearly all animals need some form of locomotion – the ability to move about. Finally, they need to be able to reproduce themselves, and how they do this is described on pages 30-31.

The act of taking in oxygen and giving off the waste gas carbon dioxide is called respiration, or breathing. Animals that live in the water get their oxygen from the water, mostly breathing through their gills which are built to extract oxygen from a liquid. Animals that do not live in water breathe through lungs if they are vertebrates – animals with backbones; and in various ways if they are invertebrates. Most invertebrates are small because they have neither backbone nor an internal skeleton to support a large body. Some, such as the land slugs and snails and spiders, have lungs, and most of the larger water-living invertebrates have gills. Most crustaceans – lobsters, crabs, woodlice and their allies – have gills, but some land-living species have developed lung-like organs, and others can use their gills for breathing air – beach fleas and some woodlice do this.

Insects have breathing organs called tracheae, which are small tubes connecting pores on the outside of the body with the internal organs. Effectively, the animals may be said to breathe through their skins.

Oxygen and food have to be transported throughout the body, and the medium by which this is done is blood. The simplest animals, such as protozoans (one-celled creatures) and sponges, have no need for blood, but most other animals have it. A substance called haemoglobin in the blood combines with the oxygen, and so enables the blood to carry it around. Haemoglobin gives blood its red colour. However, molluscs and crustaceans have blood containing haemocyanin, which does much the same work as haemoglobin but makes the blood blue. Insect blood carries very little oxygen because the tracheae take oxygen directly to the organs, and so their blood may be yellow, green or even colourless.

Locomotion depends on the size of the animal and the element in which it moves to find its food. A few kinds, such as adult sponges and corals, never move at all; they wait for their food to come to them, but their young swim about in the sea. The simplest organs used for moving in water are cilia, little

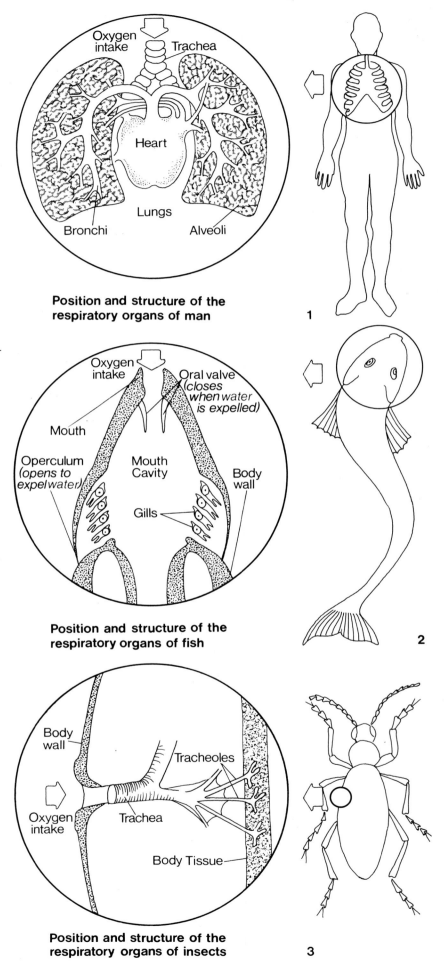

Position and structure of the respiratory organs of man 1

Position and structure of the respiratory organs of fish 2

Position and structure of the respiratory organs of insects 3

hair-like organs which wave about and 'row' the animal along. Young sponges and corals use cilia, and so do one-celled animals.

The majority of land-living animals have legs for walking. Birds go on two legs and so does Man. Other mammals, even apes and monkeys, are quadrupeds, having four legs, though apes such as chimpanzees use their front legs partly as arms. Insects have six legs, while spiders and their relatives have eight. Most reptiles have four legs, but the snakes have no legs at all, and move along by contracting and extending their bodies. Quadrupeds which spend a lot of time in water have legs which are adapted for swimming: frogs, for example, have webbed feet, with skin between their toes, while seals have paddle-shaped flippers.

Whales and most fishes propel themselves through the water by strokes of their tails, and steer by means of fins. Cuttlefishes, squids and scallops move by a kind of jet propulsion. Many very small sea animals float, and though they travel great distances they do so only because the movement of the sea takes them.

Flight is confined to birds, bats and insects. Insects have one or two pairs of wings besides their legs; birds have two legs, and wings take the place of their forelegs. For this reason insects can run about easily, but birds tend to be clumsy creatures on the ground. A few kinds of birds cannot fly, and use their wings to help them run very quickly. Penguins, which also cannot fly, swim well. The wings of bats are actually their hands, which have thin skin between very long 'fingers'. Some other animals, such as flying squirrels, glide, but cannot propel themselves through the air.

Left: The different ways in which animals breathe. **1** Man takes in air through the mouth and nose. **2** A fish also inhales through the mouth. **3** Insects have minute openings all over the body.

Right: There are a number of creatures, like this Euglena, of which it is very difficult to say whether they are plants or animals.

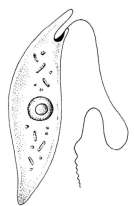

Above: The apparently frail body of a butterfly is immensely strong in relation to its weight.

Below: Animals that float in the sea, such as this moon jellyfish, use the water to support their soft bodies.

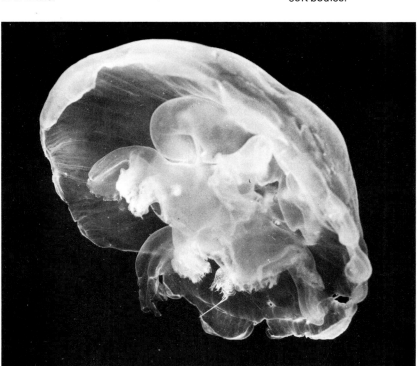

The outer casing
All animals have an outer layer which protects the soft internal organs of the body. In the higher animals this layer is called skin, and it is a highly sensitive substance. Skin is by no means a completely airtight or watertight covering: in Man and other mammals the skin is almost completely waterproof, but pores in it allow moisture – perspiration – to escape outwards as a means of losing excess heat. The surface of the skin contains thousands of nerve endings, which enable the body to feel heat, cold, pain, touch and pressure. Hair, which grows out of the skin, helps to keep the body warm and protect it from injury.

Feeding and Breeding

An animal feeds to maintain its own life, and it breeds to ensure that its species continues, that there will be more animals like itself when it is dead. These activities are both closely connected with the preservation of life.

Animals collect their food in three ways. Most of them wander about hunting for it, whether they are plant eaters or flesh eaters. A few take in the material surrounding them, soil or mud, and extract their nourishment from it. The animals that do this are those which live in the soil or in the mud and sand of the foreshore: they include earthworms, rag-worms and lugworms. The animals in the third group live by filter-feeding, and these all dwell in water. They draw in the water that surrounds them and filter out particles of food from it. Animals of this group include sea anemones, water-fleas *(Daphnia)* and a host of very small animals.

Once an animal has found its food it has to ingest it – that is, take it in. All but a few kinds of animals, such as jellyfishes and flatworms, have a through system for food on broadly similar lines. The food enters through a mouth, and passes into the gut, the digestive system. The main digestive organs are the stomach and the intestines (sometimes called the bowels). After all the nourishment has been absorbed, the left-overs pass into the rectum, the rear part of the intestines. There most of the water is removed from the un-digested material, which passes out through the anus as faeces, or waste matter. The details of the digestive system may vary, but the bodies of nearly all kinds of animals work in this way.

There are two ways in which animals repro-duce their kind. In asexual (non-sexual) repro-duction only one parent is needed to produce young. In sexual reproduction two parents, one male and one female, are involved.

Asexual reproduction is typical of the simp-lest animals, mostly those consisting of just one cell (see pages 88-89). It may consist of a simple division, as in the cells of the body (see pages 14-15). Another method, which is found in some sponges, is budding. In this method a bud, looking like a tiny copy of its parent, grows on the side of the animal, and in time breaks off to begin a new life on its own.

Sexual reproduction depends on the bring-ing together of a cell from a male animal and a cell from a female of the same species. With rare exceptions, animals of different species do not mate and cannot reproduce. As described on pages 14-15, the cells in repro-duction are special sex cells. Those produced by the female are eggs, and those from the

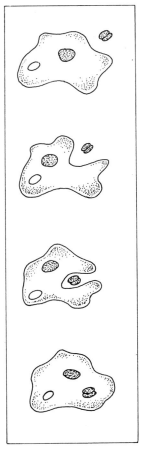

An amoeba is a protozoan. A minute, irregular blob of protoplasm, it feeds by pushing out a pseudopod – false foot – and wrapping itself around a food particle. Once the amoeba has totally surrounded the food particle, it can then absorb it.

male are sperms. When the two are brought together they unite, and a new baby animal starts to grow.

In most animals the egg passes out of the female's body and the embryo (the baby animal) develops inside the egg until it is fully formed. Then it emerges from the egg ready to begin a separate life on its own. In a few kinds of animals – nearly all the mammals (except for marsupials and monotremes), and odd species of other animals including some moss animals, reptiles, insects, scorpions and fishes – the female gives birth to live young in a fairly advanced state of development.

Fish and many other animals that live in water practise what is called external fertilisa-tion. The female releases her eggs into the water, and the male then releases his sperm. An egg is fertilised when a sperm meets it and unites with it. In such an apparently hit-and-miss method only a few eggs are fertilised, but nature makes up for this by the enormous quantities of eggs produced: for example, an ocean sunfish produces 50,000,000 eggs or more, and even a cod produces several mill-ions. The surplus eggs are eaten by other marine animals.

Internal fertilisation is the normal method for other species of animals, particularly insects, birds, reptiles and mammals. For this to take place the male mammal inserts his penis, a tube-shaped organ, into the female's vagina, a narrow, canal-shaped opening. The sperm is discharged through the penis and the vagina into the uterus, and so reaches the egg or eggs of the female. Males of other kinds of animals have somewhat different organs, but their purpose is the same: to deposit the sperm inside the female's body.

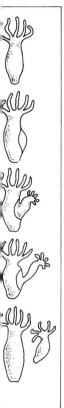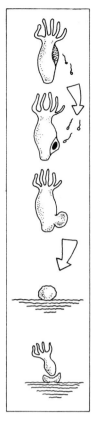

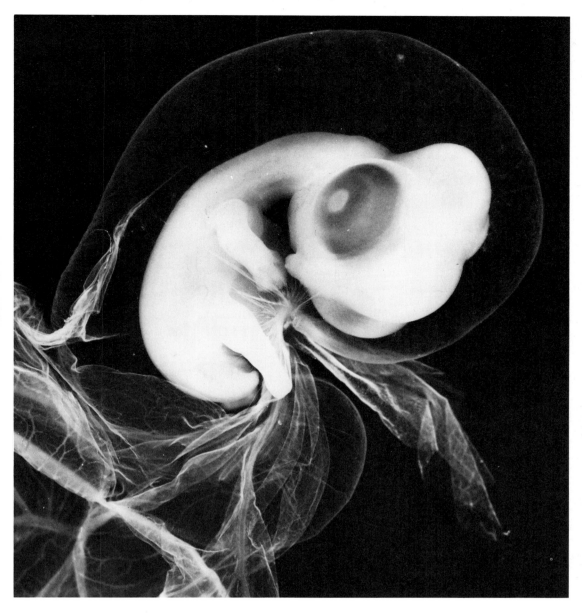

Above: Hydras can reproduce in two ways. On the left, a hydra has put out a bud, which breaks off to form a new animal. On the right, sperm from a male part fertilises an egg from the female part.

Above right: The embryo of a seven-day old chick, which has been taken from its shell.

Right: The feathery antennae of a male moon moth are used to guide the insect to a female which is emitting pheromones, chemical substances designed to attract the male.

Left: A sea anemone captures its food with its long, waving tentacles. Stinging cells on the tentacles paralyse other sea creatures and make them easy to catch.

Courtship

One of the big problems in nature is making sure that males and females meet so that they can mate.

Some animals rely on colour to attract a mate, and generally it is the male which is colourful and puts on a special display. A good example is the peacock, which flaunts his brilliant, fan-shaped tail in front of the dull-hued female.

Chemicals called pheromones, detected by the sense of smell, are exuded by many insects. Males of some moths fly long distances, lured by the scent of a female. Glow-worms flash lights, while cicadas and crickets attract one another by the sounds they make.

31

Mammals

Mammals are the most highly developed of animals in body structure and intelligence. The class of mammals is made up of around 4,000 different species, far fewer than other animals. There are, for example, about 8,500 species of birds, 25,000 different fishes, and well over a million kinds of insects. All the same, the mammals contain a great variety of species, including Man, the mighty elephant, the even mightier blue whale, the prowling tiger and the tiny harvest-mouse.

Mammals are vertebrates—that is, animals with backbones, unlike such creatures as insects and crustaceans, which have no backbones. In common with mammals, birds, reptiles, amphibians and fishes are all vertebrates, but they differ from the mammals in important ways. Reptiles, amphibians and fishes are cold-blooded: their blood-temperature alters with the temperature of their surroundings, and they do not make their own internal heat. The colder the outside temperature, the less active they become. Mammals and birds are warm-blooded, so their body-temperature stays the same, whatever the surrounding temperature. Hair or fur protection for mammals, and feathers for birds, help to maintain this constant body-temperature.

Birds and mammals differ again, not just because birds fly—for not all birds are capable of flight, and a large order of mammals, the bats, fly very skilfully. Nearly all mammals are quadrupeds—four-footed animals. Some reptiles such as crocodiles are four-footed, but the word quadruped is only used for mammals. Man walks upright, but long ago he went on all fours, just as his close relations the gorilla and chimpanzee do, though they often try to stand erect. The four limbs of sea mammals such as the seals have become flippers, almost useless on land. With mammals even more adapted to life in water, such as the whales, the front legs have been 'improved' for swimming and the hind legs have disappeared.

Another particular feature of mammals is that they all carry hair to some extent. Fur is a form of hair, whether it is the beautiful pelt of the cheetah, the velvety covering of the mole, or the bristles of the wild boar. Some animals, such as whales and porpoises, the hippopotamus and the rhinoceros, have very little hair. Only traces of it remain as reminders of far-off days when these mammals had much more hair.

A more important difference between mammals and all other animals is that they give

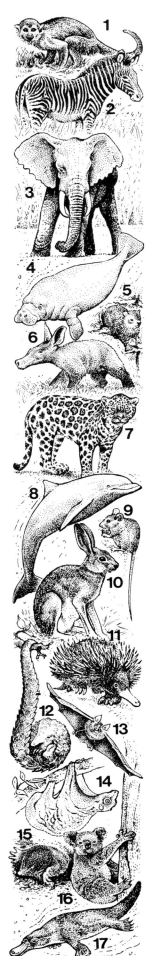

The class of mammals

Mammals are grouped together in 17 orders according to their different characteristics and ways of life:

1 **Primates,** including the apes, monkeys and Man.
2 **Ungulates,** hoofed mammals, vegetarian feeders. Zoologists split the ungulates into two orders depending on whether they have an odd or an even number of toes.
3 **Elephants,** which also used to be included with the ungulates.
4 **Sirenians** or sea-cows, which include manatees and dugongs.
5 **Hyraxes** of Ethiopia, once classed with hares and rabbits.
6 **Aardvarks,** which form an order on their own, the Tubulidentata.
7 **Carnivores** or flesh-eaters, which include cats, dogs, bears, weasels and seals.
8 **Cetaceans**—whales, dolphins, porpoises.
9 **Rodents,** gnawing animals such as rats, mice and beavers.
10 **Lagomorphs**—hares, rabbits.
11 **Edentates,** animals with weak teeth or no teeth at all.
12 **Pangolins,** scaly ant-eaters, the only scale-covered mammals.
13 **Bats,** the only flying mammals.
14 **Colugos,** gliding animals.
15 **Insectivores,** the insect-eating animals.
16 **Marsupials,** pouched mammals such as kangaroos and koalas.
17 **Monotremes,** the egg-laying mammals.

birth to living young ones, in contrast to birds and most reptiles, for example, whose young hatch from eggs.

Just to complicate matters, a few snakes such as vipers produce live young, while two mammals, the platypus and the echidna, lay eggs; but these are rare exceptions.

The most important difference of all is shown by the word mammal itself. It comes from the Latin *mammae,* which means milk-secreting organs, the mammary glands or breasts. The female mammals suckle their young with nourishing milk in the early stages, whereas birds have to fetch food for their fledglings, and many baby reptiles and amphibians have to look after themselves.

Right: The beautiful face and markings of a leopard. Leopards are usually this colour, although some are black. Leopards live in Africa and southern Asia.

Apes and Man

The primates are a group of mammals considered to be the highest and most developed order. They include the apes, that is to say the gorilla, chimpanzee, orang-utan, gibbon, and Man himself. In addition, all the monkeys and lemurs are members of the order, but various sub-divisions also exist. Although we are closely related to the apes, we did not descend from them, but we share a common ancestor who lived many millions of years ago.

The relationship between Man and ape is best seen in the behaviour of the chimpanzee. Living only in Central Africa, chimpanzees stand about 1.5 metres (5 feet) tall and go around in troops of 40 or 50. A chimpanzee searches chiefly for fruit, but it also eats insects and small mammals. Its actions are very human at times: chimpanzees embrace, kiss each other, comfort each other, hold out a hand to beg or admire another chimpanzee's baby, and they make use of very primitive tools such as sticks.

Considerably larger than the chimpanzee, the gorilla stands about 1.8 metres (6 feet) tall when upright. It prefers to travel on all fours, mainly on the ground, whereas the chimpanzee chooses to travel through the trees. Ancient legends about the gorillas' ferocity have been disproved. In general they are peace-loving animals living in small family parties seeking plant-food, of which an adult needs 25 kilogrammes (55 lb) daily. Old males sleep at the foot of trees; the others make sleeping-nests in the branches.

The orang-utan of Borneo and Sumatra is a more solitary animal. Its tree-loving habits are revealed by its powerful arms and comparatively weak legs. Slightly bigger than the chimpanzee, it is covered in tangled reddish hair and has flaps on its cheeks and throat. It has been badly persecuted, and only about 5,000 orang-utans survive in the wild.

The smallest of the apes are the gibbons of south-east Asia, of which there are seven species, all extremely active animals that live in the tree tops of the rain forest.

The monkeys are divided into two groups: the Old World species, mainly of Africa and Asia, and the New World monkeys of South and Central America. The biggest difference, apart from size, between apes and monkeys is that apes do not have tails, while all monkeys do. . The largest Old World monkey is the mandrill, 1 metre (3 feet 3 inches) tall. It also has the most highly coloured face of any mammal, its puffy cheeks being sky-blue and its nose bright red. It lives in the tropical forests of Africa, feeding on nuts, roots, fruit and small animals. Closely related to mandrills

are the baboons, which live on rocky mountainsides in groups of 200 or more. When feeding, they post sentries to keep watch against predators.

The Old World monkeys are many and varied. The rhesus monkey of India is much used for medical research. The hanuman is sacred to the Hindus. The proboscis monkey of Borneo is famous for its enormously long nose, which hangs down over its mouth. One monkey with a most unusual diet is the kra of south-eastern Asia. Its other name is crab-

Above: The ring-tailed lemur of Madagascar is one of the simplest of the primates. It is 1.2 metres (4 feet) long including the bushy tail. It spends much of its time on the ground.

eating monkey, for although like all its relations it enjoys fruit, it often goes down to the coast in search of crabs and molluscs. The only monkey existing in Europe is the Barbary ape, which lives on the Rock of Gibraltar.

The chief difference between Old World and New World monkeys is that the Old World monkeys have cheek-pouches for tucking away food when necessary, and that the tails of the New World monkeys are prehensile—able to grasp. A prehensile tail is often useful as a fifth 'hand'; the spider monkey, for example, can be seen hanging in family groups upside down from the branches. The marmosets are miniature monkeys no larger than squirrels, with bushy tails that are not prehensile and tufts of hair on their heads.

One group of primates forms a sub-order called prosimians. They are less highly developed than other primates. They include such species as the bush-baby of Africa, so-called because of its pitiful cries; it is about 300 millimetres (12 inches) long plus a long tail. Other members of the group are the graceful lemurs of Madagascar. The largest lemur is the indris, about 1 metre (3 feet 3 inches) tall, which often basks in the sun, stretched out like a human being. The lemurs are in danger of being wiped out, partly because their forest homes are being destroyed.

Above: The chimpanzee of Africa is the most intelligent of the apes. This one was photographed in the forests of Zaïre.

Left: The gorilla is the biggest and strongest of the primates. It is not as intelligent as the chimpanzee, but once it has learned something it stores the information in its memory.

Flesh-eaters

The carnivores or flesh-eaters exist in most parts of the world, and there are several families of them. The cat family includes such members as the wild cat of Scotland and other parts of Europe. Larger than the wild cat and weighing about 16 kilogrammes (35 lb) is the chunky, tufty-eared lynx of Scandinavia, Spain and Canada.

The most spectacular are the 'big cats', all of which to some degree are endangered species. Originally a northern animal, the 200-kilogramme (440-lb) tiger survives in small numbers in Siberia and Manchuria. Even in India, its chief stronghold, only some 2,000 tigers remain. The leopard, half the weight of the tiger, numbers twice that figure, while in Africa it is somewhat more numerous. In western India about 200 lions survive, but even in Africa, its chief habitat, the lion's future is in doubt.

In India the cheetah, the fastest creature on four legs—reaching 100 kph (60 mph) in short bursts while chasing antelope—is already extinct. In Africa the cheetah has been badly persecuted for its handsome skin. For the same selfish reason the ocelot of South America and the clouded leopard of south-eastern Asia have also been hunted until they are in danger of extinction. In the Americas the puma, also known as the cougar or mountain lion, weighing around 100 kilogrammes (220 lb), has been greatly reduced in numbers; so has the larger jaguar of South America, which can swim as well as it can climb. Monkeys are very scared of both these cats, which can chase them up trees.

The best-known member of the dog family is the wolf, which exists in Europe, Asia and North America. It is an extremely intelligent animal and in hunting reindeer or caribou it can separate an individual animal with the skill of a sheepdog. It does much good by concentrating on sickly or slower animals, so only the best survive. At breeding-time it catches large numbers of voles and meadow mice. The coyote or prairie wolf of North America is a small form of wolf. Far more ruthless in their pack-hunting are the dholes or red dogs of India and the hunting-dogs of Africa.

Another familiar member of the dog family is the common fox of Europe and North America. The jackal of Africa and Asia is largely a scavenger, eating the remains of kills made by lions or tigers. The dog family also includes the dingo of Australia and the strange raccoon-dog of Japan and Russia, valued for its fur and its flesh. The dog-like hyaenas are not members of the dog family. They eat carrion (dead animals) but

also kill for themselves, especially new-born antelopes.

The bear family contains the largest carnivores, the Kodiak bears of Alaska, up to 2.75 metres (9 feet) in length and 750 kilogrammes (1,650 lb) in weight. Bears such as the American and European brown bears often include fish in their diet, but in spite of their size they often eat berries and grubs. In Yellowstone Park on the borders of Wyoming, Idaho and Montana the black bear has become a scrounger, begging from tourists. The favourite prey of the magnificent polar bear of Arctic regions, which weighs up to 300 kilogrammes (660 lb), is the seal. This bear is a powerful swimmer and has sometimes been seen at sea a long distance from land. In great contrast is the Malay bear, smallest of all, about 1 metre (3 feet 3 inches) long, and with a taste for honey.

In spite of popular belief the giant panda of south-western China is not a bear. It feeds mainly on bamboo-shoots, but like its smaller relations the raccoon and the long-nosed coati of America, it catches small animals.

Among the smaller flesh-eaters are the Mustelidae family, the weasel group. They include the deep-digging badgers of Europe and North America and that expert fisherman the otter, also found in Europe and North America,

Right: A lioness feeding in the grasslands of Kenya. In a pride (family) of lions, it is the females who do the hunting, often working in groups and sharing the kill.

Below right: The timber wolf looks like a German shepherd (Alsatian) dog. It is a very cunning animal, and works well in packs to hunt its quarry down and kill it.

Below: The polar bear of the Arctic is a fierce hunter, but like all species of bears it will eat vegetable matter if no flesh is available.

where it lives along river banks. It is now rare. With its long, tapering tail, the otter can reach 1.5 metres (5 feet) in length. Its cousin the sea otter lives along the Pacific coast of North America, diving for clams and other sea-food.

The much smaller mink, closely related to the polecat and the ferret, is farmed for its beautiful fur. Like the otter it is an expert fisherman.

The stoats and weasels kill many rodents, the weasel being so small it can penetrate the rodents' tunnels. The thickset, 15-kilogramme (33-lb) wolverine of North America, Europe and Asia is nicknamed the glutton because of its greed. Well-known for a different reason is the handsome skunk of North America, which defends itself by squirting out a powerful, vile-smelling liquid.

The boldest of the flesh-eaters is the mongoose of Asia and Africa, renowned for its snake-killing feats. Its body is only about 380 millimetres (15 inches) long, and it relies for success on sheer speed; but it is also less affected by snake venom than other animals are. It is related to the very handsome genet of southern Europe and Africa.

Gnawers and Insect-eaters

In this chapter we meet several different groups of mammals which all have teeth of a peculiar kind. The first are the rodents. They are gnawing animals which vary greatly but have the same kind of teeth, the most characteristic being the front ones, known as the incisors. Hares and rabbits have two pairs of these incisors in each jaw, but the rodents have a single pair in each jaw. These teeth never stop growing. Only the front surfaces are protected by enamel, with the result that the inner side is worn away. This produces a sharp chisel-edge, which is very useful. It allows beavers, for example, to gnaw through tree-trunks, while the squirrel has no difficulty opening hazelnuts.

There is a danger in these incisors. Because they continue to grow, each set needs the matching teeth in the opposite jaw to keep it in check. If a rodent loses an incisor, the opposing tooth in the other jaw has nothing to control it and can grow to a great length, eventually preventing the animal from feeding.

While rodents are gnawing, pieces of material flake off from the object they are biting. These flakes might be swallowed but for a useful device. The rodents have no canine teeth, so there is a gap between the incisors and the molars. The hairy outer skin of the cheeks folds into this gap, dividing the mouth in two halves and preventing indigestible material from being swallowed.

The rodents are the largest group of mammals in the world. As well as familiar creatures such as rats, house-mice, wood-mice, field-voles and water-voles of Europe, they include the sousliks and chipmunks of North America and the marmots of Europe and Asia, all about 200 millimetres (8 inches) long. There is also the Old World dormouse, which hibernates (goes into a deep winter sleep) after becoming sleek and fat on nuts and insects. The tiniest of the rodents is the nest-building harvest-mouse which has a body-length of about 60 millimetres (2½ inches).

The largest rodent is the capybara which is the size of a young pig. It lives near waterways in South America and is often hunted by the jaguar. The much smaller guinea-pig is a relation. Some other rodents spend a lot of time in water, such as the musk rat and coypu of North and South America respectively. The beaver of North America and Europe, weighing up to 25 kilogrammes (55 lb), is a remarkable natural engineer. In order to create deep pools for storing its winter food of shoots and twigs, it builds dams across rivers (see pages 160-161).

The pygmy shrew of Europe and Asia is one of the smallest insect-eaters. Its body is 60 millimetres (2¼ inches) long, and it has a tail 40 millimetres (1½ inches) long. Like all shrews it feeds and rests alternately every three hours, day and night.

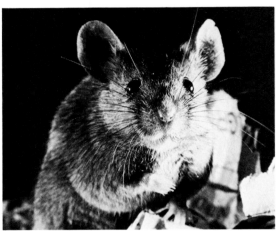

The house mouse has been Man's constant though unwelcome companion since prehistoric times. This little rodent does a great deal of damage by fouling stores of food.

The insectivores are a much smaller group of mammals which feed mainly on insects. This group includes the mole, though it feeds almost entirely on earthworms. The mole is well equipped to dig its long tunnels, for its stiff front paws are like powerful shovels. Its velvety fur can be brushed backwards or forwards, an advantage in confined spaces underground. All the insectivores have fairly sharp teeth, which enable them to catch their lively prey and crush through the hard outer cases of some insects. They include the spiky hedgehog, the tiny shrews, and strange mammals such as the moon rat of south-east Asia, largest of the group, and the rare, long-nosed solenodon of the West Indies.

The mammals known as edentates either have no teeth at all or very feeble ones. They include the armadillos of South America, remarkable for their armour-plating; and the unique, vegetarian sloths, also of South America, which hang upside down among the branches most of the time. One of the most noteworthy is the great anteater, again of South America. About 1.25 metres (4 feet) long, it has a handsome, plumy tail of the same length. Its most striking features are its very long tubular snout and its worm-like, sticky tongue, which it flicks out to pick up its chief food, termites and ants.

Mammals that fly

Bats are the only mammals that can truly fly, although several others, including the so-called flying squirrels, can glide for long distances. Bats are remarkable for the way in which they can manoeuvre at night to catch the moths on which they feed.

The bats operate a form of sonar. As they fly they send out a stream of very high-pitched squeaks. The sounds bounce off objects in their path and are detected as echoes. A few species of moths have also developed their own ultrasonic sound which they use to 'jam' the bats' sonar.

The earth-pig

If you look up any alphabetical list of animals the first one you come across is the aardvark. Its name is an Afrikaans word that means 'earth-pig'. The aardvark looks rather like a pig, but has long ears like a donkey. It is an anteater, feeding also on termites, and it lives in Africa from Ethiopia southwards. Scientists used to group it with the edentates, but now regard it as the only member of a separate order.

Above: A giant anteater out for a stroll. This one was photographed at London Zoo; in the wild these animals live in South America. They are the largest of the edentates.

Left: In its element – a Canadian beaver emerges from the water. It is one of the biggest rodents.

Vegetarians

The herbivores—which means plant-eaters or vegetarians—are ungulates, in other words hoofed animals. They are divided into two main groups, the first being the 'odd-toed'. These include all the horse-tribe, such as the few surviving wild horses of Mongolia (Przewalski's horses), the kulan and other wild asses of Asia and Africa and the zebras of Africa; the rhinoceros of Asia and Africa; and the tapir of south-eastern Asia and South America. In these animals the middle toe on each foot is the strongest; in horses it supports the whole weight.

The second group of ungulates is known as 'even-toed'. Generally speaking, these animals have four toes and the job of supporting the leg is divided between the third and fourth toes, which makes these animals cloven-hoofed. This group includes all the remaining herbivores except the elephant, which is in an order by itself.

Most of the herbivores are large, swiftly-moving animals, which run to escape danger. Many of them are ruminants (cud-chewers), including all the wild cattle, the camels, llamas, sheep, goats, antelopes and deer. These species have multiple stomachs, which enable them to eat large quantities of bulky food such as grass, and swallow it down on the spot without even chewing it. Later on they bring up cuds or wads of food, chew them and then swallow them again.

Except for the deer, most of the ruminants, especially the males, are armed with horns which they keep throughout life. The deer have antlers, not horns. These are deciduous, in other words the deer shed them annually and grow new ones. This happens even in the case of the moose or elk (of North America and Europe), the biggest of all deer at nearly 800 kilogrammes (1,760 lb) and with antlers spanning 2 metres (6 feet 6 inches). In most species of deer only the males have antlers, but both males and females of the European reindeer and its North American counterpart, the caribou, carry antlers.

Horns and antlers are used for defence. The long, almost straight horns of the gemsbok, an African antelope, have been known to kill a lion; the heavy musk-oxen of Greenland and northern Canada, when attacked by wolves, form into a defensive ring, calves and cows inside and bulls on the outside with their thickset horns lowered ready to repel the attack.

Some herbivores have horns of a different sort, notably the various rhinoceroses of Asia and Africa. The so-called 'white' rhinoceros

Sharing the food

Where wild animals are allowed to exist in their natural state, a far fuller use of resources of the area is obtained. Vegetation supports the herbivores from top to bottom, as we might say.

The tallest animals, such as elephants and giraffes, browse on the higher branches of trees. Moderately tall animals such as the gerenuk antelope, with its comparatively long neck, reach the tops of bushes. Antelopes such as eland and roan, and the little duiker and impala, rely more on the grasses. Wild pigs feed on bulbs and roots.

of Africa, largest land mammal after the elephant and weighing up to 3.5 tonnes, has two sharp, up-curving horns, sometimes 1.8 metres (6 feet) long, one behind the other. These powerful weapons are not made of horn, but consist of very densely compacted hair.

Almost as large as the white rhinoceros, and confined to Africa, is the hippopotamus, which has canine teeth more than 1 metre (3 feet 3 inches) long in the lower jaw. Its name means 'river horse', and hippos spend most of their time in water, coming on land at night to feed. The hippopotamus is very distantly related to the pig family which includes the wild boar of Europe and Asia, the wart hog of Africa, and the babirussa of Sulawesi Island,

Above: African elephants at a waterhole. Water is essential for mammals, and elephants in particular love to play and wash in it, as well as to drink it.

Above right: Thomson's gazelle is one of many kinds of antelopes which roam the African grasslands. It can reach speeds of up to 80 kph (50 mph).

Left: 'Come, let us wallow, down in the hollow' . . . a group of hippopotamuses lie almost submerged in the River Mara in Kenya.

which has a cluster of tusks curving up over its face. The only American wild pig is the fierce little peccary, which roams through the forests of South and Central America.

The largest of all the vegetarians is the elephant. With the African species weighing as much as 5 tonnes and the Asian or Indian species only a little less, these giants of the animal world require huge amounts of food. An adult elephant eats 90 kilogrammes (200 lb) of greenstuff every day, and a herd of elephants can ruin huge areas. Because of the ever-increasing human population, the survival of the elephant is in doubt. A century ago 3,000,000 elephants existed in Africa; nowadays perhaps 300,000 remain. In India today fewer than 15,000 survive.

41

Sea Mammals

Hundreds of millions of years ago the first land animals emerged from the sea and learned to breathe air. They developed legs, so they could move about on dry land. However, some mammals have reversed this process and gone back to the sea to live, although they still breathe air.

One of the largest orders of these sea mammals consists of the cetaceans—whales, porpoises and dolphins. They have completely adopted a watery way of life and never come on land. The cetaceans are divided into two groups: baleen or whalebone whales, and toothed whales.

Baleen whales are mostly very large indeed. They include the right whales, the rorquals, and the grey whales. These huge animals feed mostly on krill, tiny shrimp-like creatures. Hanging from the upper jaw of these whales is a series of flexible plates. The whales take great mouthfuls of water which they filter out through the plates, leaving the krill behind to be swallowed.

The toothed whales include dolphins and porpoises, and they have teeth to bite their food. Most of them are not as big as the baleen whales, though the largest, the sperm whale, weighs nearly 60 tonnes and is about 20 metres (65 feet) long, of which one-third is taken up by its enormous square-shaped head. These whales feed on creatures such as squids. The fiercest of the whales is the killer. Although only about 10 metres (30 feet) in length, this magpie-coloured animal is the most dreaded inhabitant of the sea and attacks even the biggest whales. The remains of 14 seals have been found in the stomach of a single killer whale.

Whales are beautifully adapted to their life in the sea. Their bodies are perfectly streamlined, devoid of hair except for a few mouth-bristles, and their strong, muscular tails drive them through the water at a speed of 15 knots (28 kph) with an up-and-down action, unlike the tail-fin of a fish, which moves from side to side. In addition whales are able to withstand the pressure of deep water much better than other mammals. The sperm whale, for example, can remain submerged for more than an hour and reach depths of nearly 1,000 metres (3,000 feet).

The pinnipeds or seals are not so completely adapted to a sea-life, having changed from being land mammals later than the whales. All the same, their limbs or flippers are of far more use for swimming than moving on land. The seals only take to the land to breed, and they spend as little time as possible there, mating again almost immediately after the pups are born. Seal-milk is the richest known, and its nutriments cause the pups to grow more rapidly than other young mammals. Common seal pups swim from birth; others such as the grey or Atlantic seal pup cannot swim for two or three weeks.

Some of the seals are large animals. The grey seal reaches nearly 300 kilogrammes (660 lb), and the Arctic walrus, famous for its moustache and ivory tusks, 1,200 kilogrammes (2,600 lb). The largest is the elephant-seal of the Antarctic, with a weight of 3 tonnes and a length of 6 metres (20 feet). The sea-lion of Californian waters is a favourite performer in circuses and marinas.

Many seals have suffered for their skins. The Alaskan fur-seal was nearly wiped out, and even today pups of the harp seal are clubbed to death in their thousands off the Canadian coast.

The sirenians form another order of sea mammals. They live in warm tropical waters grazing on sea-grasses, which gives them their other name of sea-cows. They are placid, harmless creatures. Two kinds exist, the manatees of West Africa and the Caribbean Sea and the dugong of Australia and the Red Sea. Long ago they were thought to be mermaids, and they do indeed embrace clumsily with their fore-flippers (their only limbs) and exchange a sort of kiss.

Blue whale

Sea lion

Dolphins

Sperm whale

Killer whale

Narwhal

Porpoise

Giant of the oceans

The blue whale, a kind of rorqual, is the holder of all the size records in the animal world. Not only is it the largest sea animal, but also the largest mammal and the largest living animal. Indeed no animal has ever been bigger, not even the largest of the great dinosaurs.

The biggest blue whale ever caught was more than 33 metres (109 feet) long. Most blue whales are between 21 and 27 metres (70-90 feet) long, and weigh between 56 and 125 tonnes.

The sea-otter (below)

The sea-otter lives almost entirely in the water, but is not related to any of the other sea mammals. It is a carnivore and a relation of the land otter, but its hind feet slightly resemble flippers. It hunts for crabs and shellfish off the Pacific coast of America, where it lives among the thick beds of kelp, a kind of seaweed.

It is renowned for its playful character and likes to lie on its back in the water, basking in the sun. It is also a tool-user in a simple way: it holds a flat stone on its chest and smashes open shellfish against it.

Mammals with Pouches

Most baby mammals are born in an advanced state of development, thanks to a structure known as a placenta inside the mother's womb. In these placental mammals, as they are called, the embryo—the baby growing inside the womb—takes a long time to develop; the bigger the animal the longer the time. For example, a human baby takes nine months to develop, and an elephant 22 months. A human baby is fairly helpless at first but many animals can stand and run within a few minutes of birth.

There is one group of mammals, however, whose babies do not spend a long time in the womb. These are the marsupials of Australia and America. Even for the largest marsupial, the red kangaroo, which is 1.8 metres (6 feet) tall, the gestation period (time in the womb) is less than six weeks. The baby kangaroo when born is nowhere near fully developed, and incredibly is only about the size of a bumble-bee. The only developed parts are its clawed forelimbs, with which it crawls through its mother's fur to a pouch on her belly. It is from the Latin word *marsupium,* meaning a pouch, that the marsupials take their name.

Even when the baby kangaroo gets into the pouch it is so feeble it cannot even suck for itself. The mother is equipped with a muscular device by means of which her milk is squirted into the baby's mouth. Several months pass before the young 'joey' ventures out into the open. Even when it does, it continues for another year to hop in and out of the pouch.

There are many species of kangaroos, from the big red and the forester through the wallabies to the rabbit-sized rat-kangaroo which makes a nest, for which it gathers material with its tail. All these species produce only a single baby at a time, but the quoll—the Australian native cat—sometimes gives birth to two dozen young. Her pouch is often so overcrowded that the family simply cling to her belly, even when she goes hunting through the trees for birds and rodents.

Though most marsupials are vegetarian, some, including the quoll, are carnivorous. Prominent among these is the Tasmanian devil, 1 metre (3 feet 3 inches) long and stockily built. It is a powerful creature, able to tackle prey larger than itself. It lives only in the wilds of Tasmania, also the home of the thylacine or marsupial wolf, the size of a medium-range dog, which preys on wallabies and sheep. It is not quite certain whether or not it is extinct.

Some marsupials dig extensively, such as the marsupial mole which has its pouch opening backwards, a help to an animal working in

Right: A great grey kangaroo, one of the largest of the marsupials, with a joey, a young kangaroo, in her pouch. Joeys 'thumb a ride' like this until they are about 18 months old.

Far right: A duck-billed platypus, one of the egg-laying mammals, diving in a river in search of food. When naturalists in Europe first saw the skin of a platypus they thought it was a fake.

The koala is a marsupial that looks just like a living teddy bear. It spends its life in the branches of eucalyptus trees, feeding on the leaves. The koala is a slow-moving animal, and often hangs underneath branches, just like a sloth.

the soil. Otherwise the pouch would fill with dirt as the animal moved forward. Similarly equipped are other digging marsupials such as the bandicoot, which feeds on worms and grubs, and the chunky wombat which is completely vegetarian. The wombat constructs long, deep burrows. Very different are the flying phalangers which, by means of a flight membrane between the limbs, can glide considerable distances.

The only marsupials to exist outside Australia are the phalangers of New Guinea and the opossums of North and South America. The North American opossum gave its name to the expression 'playing 'possum', because it sometimes pretends to be dead in the hope of deceiving a predator. Millions of opossums are killed for their skins and also because of their reputation for stealing poultry and eggs. The water opossum of Central America and Brazil is the only aquatic marsupial, and it catches fish extensively. The mother remains on land all the time her young ones are in the pouch. Other South American members of the family, the woolly and rat-tailed opossums, are the size of squirrels, and feed on insects and fruit. Marsupials are found as far south as Patagonia.

The egg-layers

Even odder than the marsupials are the monotremes, the only mammals which lay eggs but they do nourish their young with milk. There are two kinds. The duck-billed platypus, found only in Australia, has webbed forefeet, a sensitive, duck-like bill, and velvety fur. An expert swimmer, it hunts in rivers. The echidna looks a little like a long-snouted hedgehog and has a thread-like tongue used for catching ants. Echidnas live in Australia and New Guinea.

No competition

Although marsupials are found only in Australia, New Guinea and the Americas, fossils in the rocks show that they used to live in other parts of the world. They died out because of competition from the more successful placental mammals. Australia and South America were cut off from the rest of the world and so the marsupials developed freely there. Even when South America became linked with North America, some marsupials survived.

Birds

Birds make up a class (major group) of the animals that have backbones. They and the bats are the only backboned animals that can really fly, though some fishes, lizards and mammals can glide. However, not all birds can fly. The main features that make birds different from other animals are the shape of their bodies, and their feathers. Birds are the only animals to have feathers.

Beneath its plumage a bird is basically like the other backboned animals. Its skeleton has the same bones, but they are shorter or longer than those of other animals. The wings are made of feathers attached to long finger bones and short arm bones. The feet are long, and what looks like a bird's knee is in fact its ankle. The neck may be very long, and the jaw bones jut out from the skull to form the bird's beak. The ribs are small, but the breastbone is large. Attached to it are the flight muscles that power the wings and give the bird a plump breast.

To be able to fly, an animal has to be as light as possible in relation to its size. Birds have very light-weight bodies because their bones are hollow and their bodies contain air sacs, having the same effect as little balloons.

Most birds jump into the air or fall off a high perch to become airborne. They fly by flapping their wings and by gliding. A bird flaps its wings to push the air down and so keep itself up in the air, and to push air backwards so that it moves forwards. However, as the bird moves through the air, the flow of air over the wings also keeps it up. A gliding bird holds its wings out and moves at a slight downward angle until it finds a rising current of air to raise it aloft.

In all there are about 8,600 different species of birds. However, you will be able to see only a few kinds in any one place. These are the birds that are able to feed in that place.

Because birds can fly, many kinds have spread over the world to live. Migrating birds have two homes and fly from one to the other twice a year, often crossing oceans and continents. You can read more about migration on pages 94-95.

Birds lay eggs, which they incubate (keep warm until the baby birds hatch out). The parents feed the babies until they are old enough to find food for themselves.

Names of groups such as sea birds, birds of prey, and flightless birds may tell you where or how a bird lives or some other feature of its life. However, people who study birds place them in groups according to which birds have similar bodies. These groups are called orders, and there are 27 orders of birds in all.

Above: These drawings show how a bird moves its wings when it is flying.

Right: A baby European robin which has just fledged – acquired its feathers – and is able to fly. At this age a bird still relies on its parents to bring it food.

The orders of birds

1 Ostrich
2 Rheas
3 Cassowaries, emu
4 Kiwis
5 Tinamous
6 Penguins
7 Divers
8 Grebes
9 Albatrosses, shearwaters
10 Pelicans, gannets, cormorants
11 Herons, bitterns, storks, flamingos
12 Swans, geese, ducks
13 Birds of prey: vultures, eagles, hawks, kites, osprey, falcons
14 Grouse, partridges, pheasants, chickens, turkeys
15 Cranes, rails, moorhen, coots
16 Plovers, sandpipers, avocets, skuas, gulls, terns, auks (razorbills, guillemots, puffins)
17 Pigeons, doves
18 Parrots, cockatoos, budgerigars, macaws
19 Cuckoos, turacos
20 Owls
21 Nightjars, frogmouths
22 Swifts, hummingbirds
23 Colies
24 Trogons
25 Kingfishers, bee-eaters, hoopoes, hornbills
26 Honeyguides, toucans, woodpeckers
27 Perching birds: antbirds, lyrebirds, larks, swallows, martins, pipits, wagtails, shrikes, waxwings, dippers, wrens, mockingbirds, warblers, flycatchers, thrushes, chats, robins, tits, nuthatches, treecreepers, sunnals, finches, weavers, crows, sparrows, starlings, oxpeckers, bowerbirds, birds of paradise.

Feathers

Feathers are made of keratin, the material of which your hair and nails are made. Birds have two kinds of feathers. Contour feathers cover the body and make up the wings. They are stiff and strong. Down feathers are fluffy; they lie beneath the body feathers and keep the bird warm. Wildfowl have down beneath their outer feathers.

Sea Birds and Water Birds

Many kinds of birds live out at sea or at the seashore. Several others prefer inland waters, and live near rivers and lakes or in marshes. All these birds must come to dry land to nest—although grebes build floating nests of water plants—but they spend most of their time seeking their food in water.

It may seem strange that a creature which can fly should swim and dive to get food. However, the sea and inland waters are like a great larder, and many animals hunt fishes and other water animals or feed on water plants. Birds are light creatures and so have no problems in floating on water. Most water birds have webbed feet that they use like paddles to propel themselves through the water. Diving is not so easy, as it is difficult for such a light animal to stay underwater. Some diving birds simply plunge into the water from the air; others swim down.

Penguins live in Antarctica and the surrounding oceans. They cannot fly and they live by hunting fish underwater. Several penguins raise their young on Antarctica and nearby islands. They huddle together amid the snow and ice for warmth. Instead of nesting, some of these penguins hold their one egg on their large, webbed feet and lower a fold of warm belly flesh over it.

Divers and grebes nest inland on lakes, and divers and some grebes go to the coast for the winter. Both birds are good at diving, and spend most of their time hunting under water.

Albatrosses and shearwaters make the sea their home more than any other birds. The adults come to islands or remote shores once a year to nest, but the young may roam the oceans for years without ever coming to land.

Pelicans live on lakes and the seashore. They scoop up fish in their huge, bag-like beaks. Sometimes they form groups and go fishing together. Gannets nest in huge colonies on cliffs, and splash into the sea to catch fish. Cormorants live on rivers, lakes and also the coast. They capture fishes by chasing them under water.

Herons, bitterns, storks and flamingos mostly have long legs and long necks. They mainly live in lakes and marshes, wading in the water and lowering their heads to feed from it. Flamingos suck up water, and eat the tiny creatures that float in it. Bitterns hide away among the reeds in marshes.

Swans, geese and ducks eat plants as well as animals, and many of them come ashore to feed. Swans are the largest birds of this group, which are known as wildfowl or waterfowl. They live in lakes and rivers. Geese are medium-sized, and are found on the coast as well as in lakes and rivers, as are ducks, the smallest wildfowl. Swans and geese can lower their long necks to feed underwater. Ducks either up-end their bodies to lower their heads into the water or they dive for food.

Cranes, rails, moorhens and coots are

A group of colourful flamingos with the cone-shaped mud mounds they build to make their nests on. Flamingos live in salty, muddy water.

mostly found in lakes and marshes. Cranes are tall birds that dig in soft mud and soil for food. Rails have narrow bodies that help them to move through the reeds of marshes. Moorhens and coots are common black birds of lakes and rivers.

Plovers and sandpipers are often called waders because many wade in shallow waters, both inland and at the coast, to find food. However, many also live in woods, fields and moors, especially when they are nesting. Their group includes oystercatchers, lapwings, curlews, woodcocks and avocets.

Skuas are fierce sea birds. They live out to sea and at the shore, often robbing other birds of their food and eating eggs and young. Gulls are birds of the seashore, though several also live inland. They settle on the water to feed. Terns dive into the water from the air to feed.

Auks are a group of sea birds which includes razorbills, guillemots and puffins. They are excellent divers but, like many

'Swan flight, cleared for take-off' – a swan 'running' along the water as it becomes airborne.

A northern gannet, a goose-sized sea bird, with a beakful of seaweed as a gift for its mate.

diving birds, can walk only clumsily on land. They live in the northern oceans, and nest in large colonies on the coast. Razorbills and guillemots lay their eggs on bare rock ledges high up cliff-faces. Puffins nest in burrows.

Many kingfishers are brightly-coloured birds that make lightning dives into a lake or river to capture a fish. However, some of them, such as the kookaburra of Australia, are forest birds that do not go near water.

Dippers
Dippers are very unusual water birds for they belong to the order of perching birds which live on land. Although they do not have webbed feet like most swimming birds, dippers dive in streams to catch food. They do this by spreading their wings and facing into the current so that they are forced to the bottom. In this way, they can even walk along the bed of the stream to feed.

Hunters and Hunted on the Wing

The world of nature is a battleground. Animals that eat plants are constantly hunted by flesh-eating animals, which may in turn be captured by other hunting animals. Only a few fierce hunters have nothing to fear, except perhaps from Man. Birds take part in this eternal conflict, being either hunters or hunted, and sometimes both. Several birds such as pigeons are plant eaters, while many live both on plants and on small animals. Insects are a favourite food of many birds. Birds of prey, owls and several fierce sea birds rule the bird world, often hunting other birds as well as reptiles, amphibians and small mammals.

The hunters have several ways of attacking other creatures that help them to capture their prey. Owls, for example, have very good hearing and keen sight, so they can easily find a meal. They can also fly silently, so their victims do not hear them coming. Birds of prey have sharp claws which kill their victims, and hooked beaks which tear them to pieces.

The hunted birds are rarely able to fight back, but they do have good ways of defending themselves. Flightless birds can run from danger, and many birds simply hide. Some

A Galápagos hawk hovering. The broad wings and short neck are typical of birds of prey, and the sight of an outline like this against the sky makes ground-living birds which might be the hawk's next meal keep still so as not to be detected.

have plumage that is coloured like their surroundings, and this camouflage prevents a hunter from spotting them. Some plovers protect their nests by leading an enemy animal away from it, pretending to be hurt and so looking as though they could be easily caught. When the enemy has been led far enough, the bird simply flies away.

Flightless birds include the ostrich of Africa, the rheas of South America and the emu of Australia. They are similar, large, long-legged and long-necked birds. They live on grassy plains, and their height enables them to see danger coming. They run to escape, but may turn and kick at an enemy if they are cornered. Cassowaries are similar birds that live in the forests of tropical Australia and New Guinea. Kiwis are smaller flightless birds that live in the forests of New Zealand. They can find worms and grubs by their scent.

Although any bird that hunts another animal can be called a bird of prey, the term 'birds of prey' is normally used for a group of hook-beaked birds that hunt by day. They include vultures, which mostly eat dead animals and are useful because they clean up the remains of animals killed by other hunters. The group also includes large birds such as eagles, kites, hawks and buzzards, and the smaller falcons. These birds swoop to capture creatures on the ground or in trees, or they chase birds through the air.

Game birds get their group name because many of its members are hunted for sport by Man. Game birds include grouse, partridges and pheasants. Although they can fly, they

The ostrich is the largest living bird, found today in dry parts of central Africa. This is a female, slightly smaller than the 2.4 metres (8 feet) of the males.

50

prefer to live on the ground, where they peck for food. They are found in woods, forests, moors and fields. Many pheasants are brightly coloured, as the peacock is. Chickens and turkeys belong to this group.

Pigeons and doves mostly eat plant food. They live throughout the world in all kinds of places. Racing pigeons and the wide variety of pigeons seen in cities are all tame descendants of wild rock doves.

Parrots are brightly-coloured birds that mainly live in woods and forests throughout the tropical parts of the world. This group of birds includes cockatoos, which live in Australia and south-eastern Asia, and macaws, living in South America. The parrot feeds on fruit and grain, and is able to pick up a fruit in one claw and take a bite from it with its hooked beak. Parrots are popular cage birds, and they may learn to 'talk' in captivity. Although they can repeat words, they do not understand them.

Cuckoos mostly eat insects and live in woods. Their name comes from the 'cuckoo' call of the male common cuckoo. Cuckoos are well known because many of them do not raise their own young, but find other birds to hatch their eggs and feed their chicks. The female cuckoo lays its egg in another bird's nest and then leaves it. Even though the young cuckoo pushes out the other bird's eggs and young from the nest, it is still fed and cared for by the other bird until it is old enough to leave the nest.

Owls live in all kinds of places throughout the world. They hunt small animals, mostly by night. They use their keen hearing and sight to find their prey. Owls have eyes that face forward, like our eyes. This gives them vision in depth, making it easier for them to judge distances.

Nightjars also hunt by night, chasing insects

A pair of gold-and-blue macaws preening one another. They live in tropical rain forests from Panama south to Argentina.

The little owl of central and southern Europe is one of the smallest of the 133 species of owls. It is often seen in daylight.

through the air. They get their strange name from the jarring calls they make at night. Nightjars sleep by day, hiding on the ground or in trees. They are difficult to see because their plumage looks like bark or dead leaves.

Swifts and hummingbirds are two groups of birds that have similar bodies but live in different ways. Swifts spend most of their time in the air, hunting flying insects. They are found throughout the world. Hummingbirds live in the forests of North and South America. They are small, brightly-coloured birds that feed on plants. They can hover in front of flowers, and insert their long thin beaks into the blooms to suck up nectar.

Honeyguides feed on the wax in bees' nests, and lead animals and even people to a nest to open it for them. They live mostly in Africa. Toucans are birds of American tropical forests. They have large colourful beaks that they use to pick fruit. Woodpeckers make their homes in trees everywhere except Australia and Madagascar. They hunt insects in the bark, and carve out nesting holes in the wood, chopping away with quick, repeated blows of their chisel-like beaks.

51

Perching Birds and Songbirds

The order of perching birds and songbirds is the largest group of birds. It contains over 5,000 different species—more than half of all the birds in the world. The birds have feet with three toes in front and one behind, so they can grasp branches and perch easily. Many of them are able to sing well. All the birds in this group are land birds, and most of the birds you see in your garden and around your home belong to the group.

Lyrebirds live in the forests of Australia. They get their name because the male lyrebird has a tail that is shaped like a lyre, an old kind of harp. The lyre shape shows up when the male raises his tail in a fan of feathers to attract a female lyrebird.

Larks are brown-coloured birds that live on moors, fields and other open spaces in most countries of the world. They live on the ground, hiding among the grass. Larks fly up into the air singing when disturbed.

Swallows and martins dash through the air in pursuit of insects, and often build their mud nests on the walls of buildings. They nest in northern countries and fly south to warmer parts of the world for the winter. Sand martins nest in burrows in banks of sand or earth.

Pipits and wagtails are small birds that live in plains, moors and fields. Pipits are brownish in colour and are found throughout the world. Wagtails get their name because when they are on the ground they wag their tails up and down. They are more colourful than pipits and usually live near water. Wagtails are mostly found in Europe, Asia and Africa.

Shrikes are hunting birds. They capture large insects, frogs, lizards and small birds. Instead of eating their victims right away, shrikes often store their bodies by sticking them to thorns and—moving with the times—barbed wire. Shrikes live in Europe, Asia, Africa and North America.

Wrens are small birds that live near the ground, where they scurry about in search of insects. They carry their tails pointed upright.

Warblers live mainly in woods and forests throughout the world, seeking insects among the leaves. Many are coloured green or brown, so they are difficult to spot. Warblers get their name from their warbling songs, although many American warblers are not good songsters.

Flycatchers live in forests and woods throughout the world. As their name indicates, they hunt for flying insects, dashing among the trees in pursuit of them. Many tropical flycatchers are brightly coloured.

Thrushes make up a large family of perching birds. They live in woods and also on the

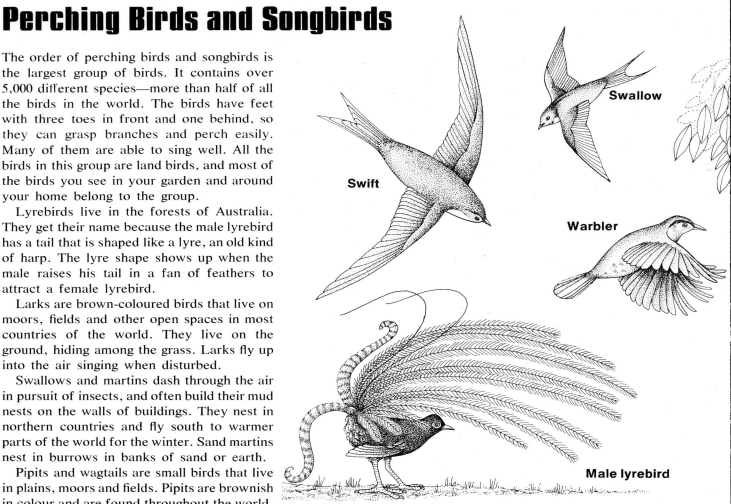

Swift

Swallow

Warbler

Male lyrebird

ground, seeking insects and also eating plant food, especially fruit. The family includes European and American robins, chats, nightingales, redstarts and the European blackbird. Thrushes are very good songsters.

Tits, or titmice to give them their full name, are small, agile birds of woods and forests. They clamber among the leaves, seeking insects, fruits and berries. Some come readily to bird tables and nest boxes. Tits are found in Europe, Asia, Africa and North America, where they include chickadees.

Nuthatches and treecreepers are woodland birds, found everywhere except South America. They live mainly on tree trunks, seeking insects in the bark. Nuthatches cling to trees head-downwards, while treecreepers run up trunks and branches.

Buntings are small seed-eating birds of woods, hedgerows and fields. In Europe, Asia and Africa, buntings are mostly brownish birds. The yellowhammer is one of the exceptions. Many American buntings have bright colours, including the cardinal, which is red, and the multi-coloured painted bunting.

Finches are small seed-eating birds of woods and fields. They are found throughout the world. Most of them are brightly coloured, including the chaffinch, goldfinch and bull-

Weaver bird and nest

Willow tit

Spotted flycatcher

Wre

Nuthatch

Tree creeper

Thrush

finch. The canary and several tropical finches are kept as cage birds.

Weavers and sparrows are small seed-eaters of Africa and Asia. However, two of them, the house sparrow and tree sparrow, have spread throughout the world alongside Man. Weavers get their name because they weave nests in unusual shapes.

Starlings mainly eat insects. They live in woods and fields in all continents except South America. Common starlings throng in city centres to sleep. Starlings are good mimics; mynah birds, which are the best 'talking' birds, belong to the starling family. Oxpeckers are African starlings that perch on the backs of cattle, feeding on the pests that infest their hides.

Bowerbirds live in the forests of Australia and New Guinea. They are famous for their courting behaviour. The male attracts a female by building a bower, a construction of walls and sometimes a roof made of sticks and twigs and often decorated with flowers.

In the forests of Australia and New Guinea live birds of paradise, whose males are among the most beautiful of all birds, with their long plumes and fans of lacy feathers. The males attract the females with their lovely colours. The females are dull-coloured birds.

A few examples of the perching birds are shown here. More than half the world's birds fall in this group, and most of them are small creatures.

Crows

Crows are among the most intelligent of birds, and most can readily adapt to any situation. Crows are found everywhere except New Zealand. They live in forests, woods, plains and fields, and they are able to eat almost any kind of food. The crow family includes rooks, jays, jackdaws, ravens and magpies.

Reptiles

Reptiles are animals with backbones and scaly skins. They all breathe air, though many of them spend most of their time in water. Unlike mammals and birds, they are cold blooded—that is, they do not make their own heat internally, but take their temperature from their surroundings. For this reason they tend to be sluggish on cold mornings. Reptiles come midway in development after fishes and amphibians and before birds and mammals. Millions of years ago, reptiles were the dominant animals in the world (see pages 166-167).

There are three basic types of reptiles. One kind has a long body and a long tail: crocodiles and lizards are of this kind. The second type has a long body but does not possess a clearly-distinguished tail: these are the snakes. The third type has a short body, and the animal is largely encased in a shell. Tortoises and their relatives are of this kind.

Most reptiles tend to live on their own, and the males and females come together only at mating time. The female always lays her eggs on land, even when she spends most of her time in the water. Some reptiles build crude nests—crocodiles and alligators do this—but most merely scrape a shallow hollow in the ground. Reptiles produce their young from eggs in three ways. Most lay eggs which incubate in a nest or on the ground, and the young eventually emerge, just like birds. Some lay their eggs only when they are ready to hatch, and the young emerge from the eggs almost at once. A few species retain the eggs inside their bodies, and the young emerge fully formed from the mother. Adders and other vipers produce young this way.

The reptiles are divided into four orders: tortoises and turtles; crocodiles and alligators; the rare tuatara; and lizards and snakes, described on pages 56-57.

Generally speaking, all the first group can be called turtles, but people often give the name tortoise to turtles that live only on land, and a few freshwater turtles are sometimes called terrapins. There are about 240 species.

Sea turtles include some of the largest species, and the biggest turtle of all, the leatherback, can grow 2.4 metres (8 feet) long and weigh 680 kilogrammes (1,500 lb). Despite their heavy shells and their bulk they are expert swimmers. The freshwater turtles spend most of their lives close to rivers and ponds, entering the water in search of food.

The land tortoises are mostly larger, the smallest being the common spur-thighed tortoise, about 300 millimetres (12 inches) long. In this group are the giant tortoises found on the Aldabra Islands in the Indian Ocean and

the Galápagos Islands off the coast of Ecuador. They have shells up to 1.5 metres (5 feet) long.

The crocodilians are divided into three groups: crocodiles, alligators and gavials. They all look much alike. The true crocodiles live in shallow waters in the tropics, favouring slow-moving rivers and swamps. There are 12 species, of which four live in the Americas. The most famous is the Nile crocodile, which is found in most parts of Africa. It grows up to 4.5 metres (15 feet) long, and feeds on fishes, birds, and small land animals that venture down to the water to drink. In Asia the best-known species is the mugger.

The biggest crocodile is the salt-water crocodile which lives around the coasts of northern Australia and southern Asia. It has been seen swimming many kilometres out to sea, as has the American crocodile which lives around the Caribbean Sea. Both crocodiles can grow up to 6 metres (20 feet) long.

Alligators are very like crocodiles, but they have broader and shorter heads. The American alligator lives in swamplands in the southern part of the United States, while the Chinese alligator is found in the lower part of the Yangtze River in China. Most of them do not grow more than 3.7 metres (12 feet) long. Caimans are closely related to alligators, and they are found in Central and South America. The largest, the black caiman, reaches a length of 4.5 metres (15 feet).

Lone survivor from the past

More than 100 million years ago reptiles in the group called Rhynchocephalia were common on Earth. Today, just one species survives. It is the tuatara (above), a lizard-like reptile that is found only on a few small islands off the coast of New Zealand. It differs from other reptiles in the structure of its skeleton, and particularly its skull. Tuataras hunt at night. They are long-lived animals, and many have been known to survive for about 80 years.

Lizards and Snakes

Lizards and snakes form one of the four orders of reptiles. Basically, the big difference between the two kinds of animals is that lizards have legs and snakes do not—though as an exception to this rule there are some species of lizards with greatly reduced limbs and one, the slow-worm, with no legs at all. Lizards have eyelids that move and they cannot open their mouths very wide; snakes have fixed eyelids and they can open their mouths to an incredible extent to swallow their prey.

There are about 2,500 species of lizards, and they are of many shapes and sizes, ranging from the Dragon of Komodo, which can be 3 metres (10 feet) long, to tiny species only centimetres in length. Only a very few lizards defend themselves with poison. Most of them trust to their speed to escape from danger, or else they bluff their way out, often by puffing themselves up and hissing. The Australian frilled lizard has a large frill of skin around its head which it can erect to make itself look much larger than it really is.

The so-called typical lizards live in Europe, Asia and Africa. They are slim, nimble animals with long tails. If a lizard loses part of its tail it grows again, though never as long as it was. Most lizards in this family are not very large, although the European green lizard can reach a length of 400 millimetres (16 inches).

The geckos are fairly small, the biggest being only about 350 millimetres (14 inches) long. It is the orange-spotted tokay of south-eastern Asia, which gets its name from the sound it makes. A gecko has large feet with minute hooks on the under surface of its toes,

which give a good grip on apparently smooth surfaces.

Iguanas and their Old World counterparts the agamids are a very mixed bunch, some living in desert places, others climbing trees. The gila monster and its cousin the beaded lizard are also desert dwellers, living in the United States, and they are the only lizards with a poisonous bite.

The giants of the lizard world are the monitors, of which the Dragon of Komodo —found on the Indonesian island of Komodo—is the largest. Monitors eat their prey whole. The 80 or so species of chameleons have another attribute: they can change colour to harmonise with their surroundings. The commonest lizards are the skinks, and some of them are easily mistaken for snakes because their limbs are so small.

The 2,600 species of snakes fall into two basic groups: those which kill their prey by squeezing it to death—the so-called constrictors—and those which kill by biting. About one-third of all snakes are poisonous. A snake can swallow an animal that is bigger in diameter than itself. Its jaws are not hinged together, but can separate and so open to an amazing degree, and the rest of the snake's body is so supple that it can stretch to accommodate its victim. The snake has numerous teeth, which are renewed from time to time. In the poisonous snakes certain of the teeth are grooved and act as poison fangs. The poison is injected as the snake bites. The purpose of the poison is to paralyse the prey, or even kill it, and so make it easier for the snake to swallow it.

The snake's forked tongue is not poisonous, but is a highly sensitive organ of touch. Many snakes, particularly vipers and some constrictors, have pits on their upper lips which are sensitive to heat. With their aid a snake can detect a victim at a considerable distance.

The constrictors are the boa and python families. Pythons live in the Old World, while boas are found in Asia, Africa and the Americas. They include the giants of the snake world, and the biggest of them are the anacondas of South America. Anacondas may grow to more than 10 metres (35 feet) long; some pythons have been known almost the same length.

The largest of the poisonous snakes is the 4.8 metre (16 feet) long king cobra of India, also known as the hamadryad. Closely related to this and other cobras are kraits and, in the Americas, the harlequin coral snake. Other poisonous snakes include the vipers of Africa, Asia and Europe, and the rattlesnakes of the Americas. The rattle of a rattlesnake is formed by a series of horny segments at the end of its tail.

Above left: One of the commonest European snakes is the grass snake, a non-poisonous species. The snake's forked tongue is clearly seen here.

Above: An Indian cobra with its head and the front part of its body erect, ready to strike. This poisonous snake causes about a quarter of all deaths from snake-bite in India.

Left: A common lizard of Europe in the process of shedding its skin. Lizards and snakes moult regularly.

Poison record
The most poisonous snake in the world is said to be the tiger snake of Australia. Its poison glands hold enough venom to kill more than 100 sheep. Just as poisonous is the Australian taipan. Both these snakes are related to the cobras of India, but they are many times more dangerous. Unlike other parts of the world, Australia has more poisonous snakes than harmless ones.

Sea serpents
Fifty species of snake have taken to the sea, and most of them rarely come on to land. Although they breathe air like other reptiles, they can stay underwater for up to eight hours. They are poisonous, but few of them are dangerous to swimmers. All these sea snakes are small or medium sized—not to be compared with the huge sea serpents of legend, no traces of which have ever been found.

Amphibians

Amphibians, animals which divide their lives between land and fresh water, are the direct descendants of the first animals which emerged from the ocean depths into the air and sunshine, millions of years ago. Although their ancestors came from the sea, amphibians are no longer adapted to living in salt water. The average amphibian hatches from eggs laid in a pond or stream, and spends the first part of its life in a larval (immature) form in the water. As an adult it comes to land, but returns to the water to breed. However, as with everything else in nature, there are exceptions to this general rule.

There are three groups of amphibians: anurans—frogs and toads, which have no tails in the adult state; caudates—newts and salamanders, which have tails; and caecilians —which have no legs. Although these three groups differ so much, all amphibians have some characteristics in common. For one thing, they have soft, moist skin, unlike the dry skin of reptiles. Many of them can change colour to match their background. Most amphibians can exude a form of poison from their skins—a defence against any animal that tries to eat them, though this does not always save their lives. Like other vertebrates, amphibians have skeletons, though these skeletons are partly made of cartilage (gristle) rather than of bone.

There are 15 families of frogs and toads. It is not always easy to say what is a frog and what is a toad, but one of the 15 families consists of true frogs (Ranidae) and one family contains the true toads (Bufonidae). True frogs live in or near water and have smooth skin, while true toads have much drier skin, covered with warts, and live on land, going to the water only to mate and lay their eggs. Members of the other families contain members that are like true toads or true frogs, though their popular names can be misleading: for example, spadefoot toads are really frogs.

Frogs and toads have long, powerful back legs which enable them to leap great distances, and also to swim well. Most frogs have good eyesight, hearing and sense of touch. Nearly all frogs have strong, raucous voices, which they use to attract females during the mating season. The larval forms of frogs and toads are called tadpoles.

The true toads live everywhere except Antarctica. They have been introduced to Australia by Man. They mostly like drier habitats than frogs, but must find a pond at breeding time.

True frogs are also found everywhere except Australia and Antarctica. There are

The common frog of Europe is a typical member of the true frogs (genus *Rana*). True frogs are found in all parts of the world except Australia and the Pacific islands, though some have been taken to New Zealand.

many different species, some poisonous, some not. Tree frogs, members of a different family, have sticky discs on their feet which help them to climb trees. There they feed on insects, which they capture and swallow while leaping.

Nearly all newts and salamanders have a tail and four legs, though a few species are legless. Like most frogs and toads they return to their 'home' ponds to breed, often travelling long distances to do so. They spend a great deal of their lives on land, and some species even pass through the larval form away from water. As with frogs and toads, the names 'newt' and 'salamander' can be confusing. Newts are a form of salamander in which the males' tails and back fins grow larger during the breeding season. In their land form newts are sometimes called efts.

A particularly interesting form of salamander is the mudpuppy of North America. The adult has lungs but also retains the water-breathing gills of its youth—and it can absorb oxygen from the air through its skin.

The Peter Pan amphibian

Peter Pan, in the James Barrie story, was a little boy who never grew up. One Mexican species of amphibians is a Peter Pan type—the axolotl (above). The adult axolotl retains all the features of an immature animal, looking just like a larva; but it is capable of breeding.

For many years scientists thought the axolotl was a separate species, but then they found that it was what it appeared to be—a larval salamander. It retains its juvenile shape because it does not get enough iodine, and so its thyroid gland does not trigger off the change to adulthood. Axolotls fed on thyroid extract complete the change and become ordinary salamanders.

A female great crested newt laying an egg and wrapping a leaf around it. Newts are found in water during the mating and egg-laying season, but not so much at other times.

A pair of toads in mating position, with ropes of spawn (eggs) under water.

Footless amphibians

The group of amphibians known as caecilians lives in the tropics. The animals are wormlike in shape, with no legs, and although there are at least 50 species nobody knows very much about them. Some caecilians live entirely in the water.

Fish

Fishes are animals with backbones that first appeared in the ancient oceans of the Earth about 400 million years ago. They breathe oxygen, but they obtain it from the water in which they live. Instead of lungs, almost all fishes have gills, organs that can extract the oxygen from the water.

Fishes can be divided into two main groups: those such as sharks and rays whose skeletons are made of cartilage (gristle); and those that have bony skeletons, such as goldfish, tuna, sticklebacks and herring. The long backbone, whether made of cartilage or bone, acts as a support for the body muscles. These muscles, as they work, flex the body so that the tail beats from side to side. This action propels the fish through the water. The fins were originally used like keels to steady the body. The shark's pectorals, the first pair of fins on the underside, act rather like the wings of an aeroplane and provide lift to keep the shark up in the water. Most sharks sink if they stop swimming because they are heavier than water. Some of the deep-sea sharks are very oily and the oil buoys them up. Inside many bony fishes is a bag full of gas called a swimbladder, which makes them buoyant. The fins are then no longer needed to plane the fish up and they have become modified to act like little propellers, so the fish can perform very delicate manoeuvres.

Some types of bony fish use their swimbladders as hearing aids. A minnow from your local stream can hear much the same range of tones as humans can. In contrast a herring is deaf to all noise other than the lowest of notes. However, most fishes are very sensitive to vibrations in the water and can detect the direction from which the movement is coming. They also have a superb sense of taste or smell: a migratory salmon smells out its home headstream when returning to spawn (lay its eggs).

Fishes are mostly cold-blooded, like reptiles, but some of the fast-swimming predatory species such as tuna, tiger sharks and mackerel have a certain kind of red muscle. This muscle retains the heat it generates while working, and by warming up nearly 10°C (18°F) above its surroundings, it becomes three times as powerful as the other muscles.

Fishes mostly breed by the females releasing their eggs into the water at the same time as the males release milt (sperm) to fertilise them. In many sharks the male has the anal fins modified into claspers which are used to fertilise the eggs inside the female. She then usually lays each of her large yolky eggs inside a tough egg-case.

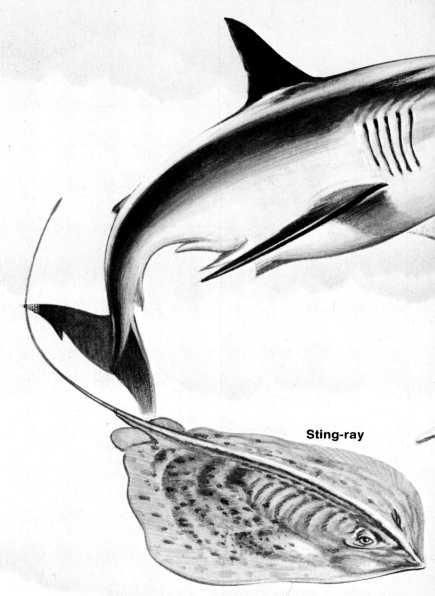

Sting-ray

A representative group of the cartilaginous fishes – sharks, rays and their relatives the chimaeras (also called ratfish). There are about 600 species all told.

Sharks and rays

The best-known of the cartilaginous fishes are the sharks and rays. Sharks, those big, ferocious hunters with their vicious teeth, frighten both sailors and swimmers. Yet the biggest shark, the whale shark, which grows to over 17 metres (55 feet) long and can weigh 50 tonnes, is a gentle giant, filtering plankton out of the water.

Manta rays that can grow to almost 5 metres (16 feet) across their great wing-like bodies are also plankton feeders. The flattened bodies of rays are adapted to life on the sea bed. Stingrays eat shellfish and have teeth suited to grinding up the shells. The sting is at the base of the fish's tail.

White shark

Mako shark

Tiger shark

Thresher shark

Bull shark

Hammerhead shark

61

Bony Fishes

The bony fishes live in almost every type of water. They have been photographed on the sea bed at a depth of 8,500 metres (28,000 feet). The ice fish lives among the pack ice in the Antarctic. It has a sort of natural anti-freeze to stop its blood from turning into ice. The blackfish of Alaska and Siberia can survive being entombed in solid ice for many weeks; whereas a small cichlid fish lives happily in the hot soda springs of Lake Magadi in Kenya where the water temperature is 45°C (113°F). Some fish can withstand being dried up for months at a time; the African lungfish survives in a cocoon of mucus, buried in the mud of its pool as the water evaporates in the hot season. At the other extreme, the mudskipper that lives in mangrove swamps is so used to breathing air that it drowns if it is held under water.

Together with this remarkable ability to live almost anywhere in water, fishes show an amazing range of body shape, colour and life style. The basic fish shape is herring-like with a streamlined body ending in a tail fin. On the back is a dorsal fin; there are two pairs of fins, the pectorals and the pelvics, on the underside, and another single fin, the anal, near the tail. Many fishes have darkly-coloured backs and pale bellies. This is a form of camouflage, and in the upside-down catfish, the colour pattern is reversed. In the deep sea at depths of around 500 metres (1,600 feet) fishes are camouflaged by having mirror-like sides. They also have little light-emitting organs along their bellies, which break up the outline of their silhouettes to predators hunting them from below. Really deep down in the ocean, fish tend to be all black.

In shallow water, camouflage is most important. The plaice can change its colour to fit the shade of the sea bed where it is lying. The sargassum fish is the same colour and the same texture as the floating weed (sargassum) among which it lurks. In fresh water the striped sides of pike and perch make these lurking predators hard to see.

Fish sometimes need to be seen, and at breeding times they develop highly-coloured fins which attract their mates. Male Siamese fighting fish develop their bright colours as a threat display to rival males. The brightly-coloured fishes found around coral reefs defend territories with colour displays, in the same way as songbirds defend territory in gardens with song. The young, immature stages of these gaudy fish are drab and colourless and do not fight for territory.

Anemone fish advertise themselves as being nasty; predators are deterred from attacking

them by the stings of the huge anemones among whose tentacles they dance. Cleaner-fish dance and advertise their presence; this behaviour persuades other larger fish to pose and be cleared of parasites.

The bodies of fishes are highly modified for their way of life. Fishes living on the bottom have become eel-like, or flattened like the plaice. In the deep sea where food is very scarce the fish are all mouth and stomach. They cannot afford to be big and muscular so they sit and wait rather than hunt their prey actively. The seahorse has its body enclosed in a jointed box of scales; its tail is prehensile (capable of grasping) and so it can cling to seaweeds or coral, and it swims along by sculling with its fins.

Above far left: The grayling is a freshwater fish of Europe and North America.

Below far left: A flounder, a salt-water flatfish.

Above left: The seahorse swims around in an upright position. The male carries the eggs in a pouch until they hatch.

Below left: The mouth of a lamprey (see below).

All-purpose fins
Fish use their fins for all sorts of purposes. The pelvic and pectoral fins of flying fishes can lock spread out, so the fish use them to glide over the waves. The freshwater hatchet fish uses its huge pectoral fins to leap out of the water. It skates over the water surface like a power-boat, with just its keel-like belly breaking the surface.

On the sea-bed gurnards use the rays of their pectoral fins like fingers for 'walking' over the bottom, tasting and feeling for their food. The spines of the dorsal fin in sticklebacks are weapons, and in weaverfish and lionfish the fin rays are tipped with poison.

Jawless fish
The early fishes of 400 million years ago were jawless, and their closest descendants, the lampreys, still occur in our streams and rivers today. The mouth of a lamprey is a disc (see left) with which it sucks on and rasps away at its prey. Behind the mouth are seven gill openings. There is a system of valves which allows the lamprey to stay sucked on to its prey while still pumping water through the gill chambers to extract oxygen.

Animals with Spiny Skins

One of the most exciting animals to find on the seashore is a starfish. Starfish belong to a group of animals called echinoderms, which all have a basic body design of five arms symmetrically arranged round a central mouth. Echinoderms only occur in the sea, and most live on the sea bed or on the seashore. There are five main types—the featherstars, the brittlestars, the starfish, the sea urchins and the sea cucumbers.

The featherstars are the most ancient group. Their fossils occur in rocks which are 550 million years old. Their skeletons of limey plates preserve well as fossils. The stalked featherstars have their bodies on the end of a long stalk which is anchored to the sea bed. Five thin arms are arranged around the body. In some types the arms divide once or even several times and on each branch of the arm occur numerous little side arms. Along the centre of each arm and side arm is a groove which is flanked on either side by tube feet. These suckered feet are water-filled tubes of skin that can be moved and squeezed with muscles, and so extended or drawn in by water pumped in and out. The featherstars use tube feet for collecting food, but some other animals move along with their tube feet. The food that the tube feet collect is passed along the food groove to the mouth in the centre of the top of the body. One common featherstar, the rosy featherstar, has a group of movable claws instead of a stalk. These claws can either anchor the animal to a rock, or move it to a new position.

Imagine the claws or stalk removed from a featherstar lying mouth downwards, and you have the basic pattern of a brittlestar. Brittlestars row themselves over the sea bed by writhing their arms. Most of them feed on fine mud particles, or on small animals they encounter on the sea bottom.

Thicken up the arms of a brittlestar and make them more rigid, and you get the starfish pattern. The tube feet, which are suckered, pull the animal along. Starfish are carnivores and many of them eat a lot of bivalve (twin-shelled) shellfish. To do this a starfish grasps both shells of the bivalve with its tube feet and pulls them very slightly apart. It then extends its stomach out of its mouth in through the gap between the shells, and digests the shellfish still inside the shells.

Imagine now a starfish with its five arms arched up over its back and joined along their edge so that the skeleton forms a hollow bell; you have a sea urchin. The mouth is on the underside, and out of it projects a ring of teeth with which the urchin rasps away at seaweed.

Over the outside are arranged movable spines that may be long, thin and tipped with poison, as in tropical reef urchins, or thick and heavy, as in the slate-pencil urchin. Heart urchins burrow into sand and have an oval shell which, when washed up on the shore, is known as a sea potato. Sand dollars are very flattened, and may have slits or even holes through their disc-like bodies.

Sea cucumbers are like sea urchins that have been laid on one side, so that the mouth points forward, and rolled out into a sausage shape. The hard skeleton is reduced to tiny plates and anchor-shaped spines. So sea cucumbers look much more like big fat worms, and indeed some live like worms, crawling over or burrowing through the sea bed eating mud.

Ancestral link

Echinoderms breed by releasing many thousands of eggs into the water. The eggs hatch into tiny larvae. The form of these larvae shows that sea urchins are very distant relatives of sea squirts, and are ancestors of the vertebrates. Sea squirts are closely related to the vertebrates, and have a structure called a notochord which represents a very primitive kind of backbone, though not a true one.

Right: The edible sea urchin lives on the seabed just offshore, though it sometimes comes inshore.

Below: A common starfish lying on a bed of mussels, which form part of its regular diet. It also eats crabs and gastropods.

Below right: A sea cucumber using its tube feet to cling to red algae on a rock.

Below far right: A brittlestar gets its name because if you handle one with anything but the greatest care its arms may break off. Fortunately the animal can grow new ones.

Molluscs

The molluscs are an abundant and widespread group of invertebrate animals in the sea and in fresh water. The only molluscs able to live on land are slugs and snails. Snails have a head on which they have stalked eyes that retract if they touch anything unpleasant. A snail glides along on its large flat foot leaving behind a sticky mucus trail. On its back it carries its shell, which is formed at the edge of a fold of tissue called the mantle. With land snails the inside of the mantle acts as a lung, but most marine snails have gills inside them. At the first hint of danger a snail can draw itself inside the shell with a muscle that is attached to the end of the spine. Some snails even have a plate called the operculum, which shuts the opening of the shell when the snail is inside.

Limpets are snails that live permanently inside their shells, and the shell is cone-shaped. The foot acts like a huge sucker to clamp the shell to a seashore rock if danger threatens. The shell also traps sea water inside so that the limpet does not dry out when the tide goes out. There are limpet-like snails called coat-of-mail shells which have eight plates to their shells. One extraordinary snail that was trawled up from the deep sea in 1952 was found to be partly segmented like a polychaete worm (see pages 68-69). Its fossilised shell was already known from rocks 350 million years old.

Snails with either tiny shells or else no shell at all are called slugs. On land, slugs are usually drab, slimy animals that live in damp places, but in the sea some of the most colourful creatures are the sea slugs. Their bright, gaudy colours are often warnings to other animals not to touch them. Many types eat sea anemones, and can use the stinging cells of the anemones they eat for their own defence. Cone shells are the molluscs most dangerous to man. They shoot poisoned darts into fish which they eat, and the darts of some species can kill a man.

Bivalve molluscs, such as clams, have two

Above: The common or garden snail is a typical example of a land snail. This kind is found in Europe.

Right: The grey sea slug is one of the larger species of sea slugs, and some are more than 90 millimetres (3½ inches) in length.

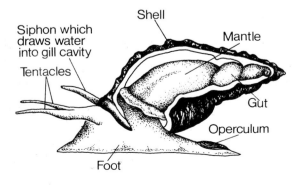

Siphon which draws water into gill cavity

Tentacles

Shell

Mantle

Gut

Operculum

Foot

Left: The inside structure of a sea snail.

valves or shells, which enclose the body. They have no head, and the foot is used either for digging, as in the cockle, or for attaching anchoring thread to rocks, as in mussels. Bivalves collect food by filtering fine particles from the water that they pump through their finely-meshed gills. There are many bivalves which are used by man. Oysters are grown not only for food, but also for producing pearls. In northern Spain mussels are grown attached to ropes hung from boats moored in the sheltered rias (estuaries). Scallops, which are trawled up, can swim by snapping their two shells together, so squirting water out in one direction. In this way scallops move by a form of jet propulsion.

Perhaps the most fascinating molluscs are the cephalopods, which include octopuses, squid and cuttlefish. Octopuses are the cleverest of all invertebrates. Their eyes are as good as ours, they have good memories and they can distinguish between quite complicated patterns. In the Mediterranean Sea there are 'cities' where octopuses live, each with its own 'home' which is a collection of stones. In Australian waters the blue-ringed octopus, only a few centimetres long, causes several deaths each year with its poisonous bite.

Octopuses move along the bottom using their eight suckered arms, or they can swim by squirting water from their mantles. Cuttlefish also jet-propel themselves through the water, but they use the fins on the sides of their bodies for delicate movements. Cuttlefish are masters of disguise: patterns of colour can ebb and flow over their bodies. They live at quite shallow depths, feeding on shrimps which they flush from their hiding places with jets of water.

Most mysterious are the squid, many of which are torpedo-shaped, high-speed swimmers. In the Peru current shoals of 2-metre (6½ feet) long squids are as fearsome a pack of hunters as any animals on land. Squid are the main food of sperm whales. The whales even eat the giant squid, large species of which exceed 12 metres (40 feet) in length.

Above: The curled octopus lives among rocks in the northern seas.

Right: A close-up view of the suckers on a squid's feeding arm. The hooks around the edges help the squid to grip its prey.

Below: A giant clam, which lives on coral reefs near Australia and the islands of south-eastern Asia.

Guide to molluscs

Monoplacophora Little-known marine animals, a bit like limpets.

Amphineura Sea molluscs, including chitons.

Gastropoda The biggest class of molluscs, including slugs, snails and limpets.

Scaphopoda Tusk shells, which live in very deep water.

Bivalvia Molluscs with two-part shells, including oysters, cockles, clams, mussels, razor-shells.

Cephalopoda Squid, cuttlefishes, octopuses.

Worms of All Kinds

Worms are animals without backbones that have long thin bodies. They live in mud or in among tangled roots or leaves, or inside the bodies of other animals—in fact anywhere where access is impossible unless an animal can squirm its way in. Thus many of the types of worm are totally unrelated, and only have in common the similar forms of their bodies and their ways of life.

The most familiar worms to the majority of people are earthworms. They are common in soils which are sufficiently moist and contain some humus (decayed matter). An earthworm burrows through soil by using a series of contractions (shortening and extending its segmented body) to ram its way through. It has stiff bristles which it can stick out sideways to provide a grip for the push. Earthworms greatly increase the fertility of soil by aerating it, improving the drainage and churning it up.

Closely related to earthworms are leeches. Leeches have a sucker at each end of the body. They move either by looping along, or by swimming with the waves of contractions travelling vertically along the body rather than from side to side as in eels.

In the sea leeches are rare, but in lakes and rivers they are very common. They are all carnivores, with different sorts eating either insect larvae or snails, or sucking the blood of mammals. In tropical rain forests, where it is always very humid, there are land leeches that attack man. Blood-letting using leeches was widely practised as a form of medical treatment before modern medicine was developed.

Other relatives of the earthworm are the polychaetes or bristle worms. They are all marine and include the lugworms and ragworms which fishermen use for bait. Tube worms, which live in sand or in limy tubes on rocks, extend beautiful fans with which they filter fine pieces of food out of the water. Palolo worms, which occur in some tropical seas, swarm at the sea surface in vast numbers to breed during full moon. The swarming worms attract birds and shoals of fish which feed on them. In the East Indies, people collect them to eat. Each segment of a palolo worm is extended at the side into a paddle. When an eel swims the S-shaped waves which push it through the water travel down the body from the head to the tail. When a palolo worm swims the waves move forward from the tail to the head—but it still goes forwards.

There are several groups of worms that have unsegmented bodies. Flatworms live in water or in constantly damp places on land. The most brightly-coloured varieties occur on

Right: One of the many species of earthworm. These worms are found in moist soil in many parts of the world.

Below: A fan worm, one of several kinds of worms living in the tubes which they make under water.

coral reefs. They glide over the ground using thousands of tiny cilia, little hair-like structures, to row them along on a mucus sheet. Many are carnivores, using their probosces to attack other animals as big as themselves. Liver flukes and tapeworms are parasites closely related to flatworms. The widespread occurrence of these parasites inside the bodies of animals such as pigs is one reason why we have to be careful how we prepare our meat and make sure it is properly cooked.

Some roundworms are also parasites. Hookworms are widespread in some tropical countries. The larvae burrow into people's skins, and are carried in the blood stream to the lungs. From there they are coughed up and swallowed. They burrow into the gut wall and live there. Many roundworms are parasites on plants. The eel worm is an important pest of potatoes, but there are worms that live freely in soil where they help to keep it fertile.

Arrow worms are little transparent planktonic animals that occur only in the sea. Beard worms burrow into mud on the sea bed, and horsehair worms spend the first half of their lives as parasites and the second half as free-living carnivores. The record for length goes to the bootlace worm, which is a nemertine or proboscis worm. One specimen measured over 50 metres (160 feet) long.

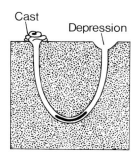

Cast Depression

Top: A mass of lugworm casts on a sandy beach. Lugworms are known as lobworms by some people.

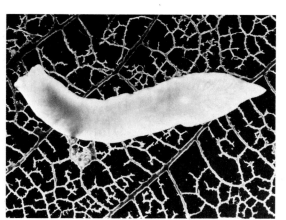

Left: A freshwater flatworm crawling over the skeleton of a leaf under water. This milky-white animal feeds on smaller animals.

Left: The acorn-shaped head and part of the body of an acorn worm, which lives in gravel and sand on the lower part of the seashore.

Digging for lugworms

Lugworms live in U-shaped burrows in muddy-sand beaches at about mid-tide level. At one end of the burrow is a depression, at the other a cast. Dig down 250 millimetres (10 inches) between the two ends with a fork to find the juicy worm, ideal bait for fishing. (Left.)

Guide to worms

Worms of various kinds are found in a number of phyla, including the following:

Mesozoa Tiny parasites inside other invertebrates.

Platyhelminthes Mostly parasites in larger animals; they include tapeworms.

Nemertinea Ribbon worms, mostly living in the sea.

Acanthocephala Parasitic worms with spiny heads.

Phoronidea Worm-like animals that live in mud under the sea.

Sipunculoidea Peanut worms, which also live in mud on the seashore.

Annelida The segmented worms, including earthworms and leeches.

Chaetognatha Arrow worms which form part of the plankton.

Pogonophora Beard worms, also living in the sea.

Hemichordata Tongue worms, living in mud under the sea.

Animals with Jointed Legs

If sheer numbers were a guide to importance, then the group of animals known as arthropods would be the rulers of the animal kingdom, for there are well over 1,100,000 known species, more than three-quarters of all the different kinds of animals in the world. Arthropods are the most successful of all the invertebrates, and members of the phylum live in almost every imaginable habitat. The name arthropod comes from Greek words meaning 'joint-footed', but it is not a strictly correct name because arthropods in fact have jointed legs, not jointed feet.

There are ten classes of arthropods, of which the most important are the insects; the arachnids (spiders and their allies); the crustaceans (including crabs, lobsters and wood-lice); centipedes; and millipedes. Scientists think that, many millions of years ago, the arthropods evolved from the same ancestors as the annelids—the segmented worms. One arthropod class, the velvet worms, is possibly a 'missing link' between annelids and arthropods. Velvet worms look something like slugs with legs, and externally they are more like annelids, while their internal organs are more like those of arthropods. The best-known genus is *Peripatus*, by which name velvet worms are often known.

Almost all arthropods have a hard, horny outer casing known as an exoskeleton, or external skeleton. It protects the animal and serves something of the same purpose as a true skeleton in that it forms a support for the body. Having an exoskeleton is rather like living permanently inside a suit of armour, and when an arthropod is growing it has to moult, or shed its exoskeleton. Soon after moulting the soft outer layer underneath hardens to form a new, larger suit of armour.

The bodies of arthropods are made in sections, and the exoskeleton is jointed to allow for ease of movement. Each section has a pair of limbs for walking or swimming, but in the course of evolution some of these pairs of limbs have changed their function or disappeared. For example, in many species one pair of limbs at the front has become a pair of antennae, or feelers. Most arthropods have eyes, and some of these eyes are compound—that is, they are made up of many lenses.

Arthropods were among the first groups of animals to evolve. The earliest fossils (stony remains) that have been found are in rocks that are 600,000,000 years old. These fossils are those of trilobites, little sea animals with shells divided into three lobes (sections)—hence the name trilobite.

Top: A centipede is a flesh-eating hunter.

Centre: A millipede is a vegetarian, and some species eat the roots of plants, and also tubers such as potatoes.

Bottom: *Peripatus* is an animal which is in many ways like an annelid worm. Some scientists do not even include it among the arthropods.

Opposite: The water spider lives in ponds and other stretches of fresh water, and spends its whole life submerged. It lives in a dome of silk which it fills with air. It brings the air down in bubbles from the surface.

Many-legged animals

Two groups of arthropods are often referred to together as myriapods—animals with countless feet. Although they look similar, having long bodies and a great many legs, they are really two distinct kinds: millipedes and centipedes.

The name millipede comes from two Greek words meaning 'thousand legs', but even the longest millipede has only about 750 legs, and most of them have even fewer. They range in size from 2 millimetres (2/25th inch) to 300 millimetres (12 inches), and there are more than 7,500 species. The millipedes are vegetarians, living mostly on decaying plants. However, many of them attack living plants and they can be pests on farms and in gardens. They walk by contracting a group of legs together, then spreading them out again with a rippling motion.

Centipede means 'hundred legs', yet some centipedes have as many as 350 legs, while others have only 14. There are 1,500 species, ranging in length from 25 millimetres (1 inch) to 270 millimetres (10½ inches). Centipedes walk by alternately moving the legs opposite to each other. They are ferocious carnivores, eating insects and other small animals, and even each other. They hunt by night, and use one pair of legs as claws.

Guide to arthropods

Onychophora The velvet worms, including *Peripatus*.

Pauropoda Tiny animals, none more than 1 millimetre long, living in damp places.

Diplopoda The millipedes.

Chilopoda The centipedes.

Symphyla Small white animals up to 8 millimetres long, living under stones and dead leaves.

Insecta The insects, including beetles, butterflies, bees, flies and bugs.

Crustacea Lobsters, shrimps, crabs, woodlice and barnacles.

Arachnida Spiders, scorpions, ticks, mites and harvestmen.

Merostoma Horseshoe crabs.

Pycnogonida Sea spiders.

Crustaceans

Crustaceans are a varied and widespread group that live in the sea and fresh water, but few inhabit the land. They are enclosed in hard external skeletons impregnated with lime, so before they can grow any bigger they have to moult. After they have shed their outer casing, crustaceans are very vulnerable until the new skeleton has expanded and hardened, so they hide away where their enemies cannot find them.

Crustaceans range in size from tiny planktonic animals, such as water fleas and copepods a millimetre or so long and even more minute animals that live in the spaces between sand grains on the sea bed, to giants such as the Japanese spider crab whose legs can span over 3 metres (10 feet) and the coconut crab that climbs palm trees to pick the coconuts.

The jointed legs of crustaceans have proved to be a design that is highly adaptable to all sorts of activities. For example, crabs have four pairs of walking legs, of which the last pair is flattened into paddles in swimming crabs. The pair of pincers are legs modified for manipulating food and as weapons against attackers. Some male crabs have one pincer greatly enlarged. Male fiddler crabs use their enlarged pincer as a semaphore flag to attract the females. Under their abdomens crabs have a series of small legs on which the female carries her eggs when she is 'in berry', until they hatch. Prawns use these legs for swimming and as gills. Water fleas have flattened legs, fringed with hairs that are used as sieves to filter the fine particles they feed on from the water around them.

Crustaceans have a wide range of breeding habits. Water fleas usually lay eggs that do not need to be fertilised; this kind of reproduction is called parthenogenesis. The female broods the eggs in a pouch inside the valves of her shell until the young are mature enough to fend for themselves. Fertilised eggs are enclosed in a special hardened case that can survive being dried up.

Fairy shrimps and brine shrimps have eggs which need to be dried for several weeks and re-wetted before they are able to hatch, so these are animals that specialise in living in temporary ponds. Barnacles, like many other marine crustaceans, have planktonic larvae. Zoologists in the early 1800s thought barnacles were molluscs because living inside their limy plates they look like little limpets. Eventually their larvae were shown to be like those of copepods and crabs.

Crustaceans form interesting associations with other animals. Sandhoppers are familiar

Woodlice are the most readily noticed of the land-living crustaceans. They live in damp places from the Equator to the polar regions. The ones in this picture are on the underside of a piece of bark, a favourite haunt.

An acorn barnacle feeding under water. It uses its feathered limbs, called cirri, to filter tiny organisms out of the sea water around it.

animals on the shoreline, feeding on decaying seaweed thrown up by the tide. They have relatives in the deep sea whose larvae live inside jellyfish. Goose barnacles are occasionally washed up on the beach attached to flotsam, but one species settles on the skins of whales. Pea crabs live inside the shells of mussels, taking food off the mussels' gills.

Woodlice are crustaceans which are totally terrestrial (land-living). One woodlouse species lives as a guest inside ants' nests. Perhaps one of the most remarkable relationships is between the cleaner shrimp and its clients. This gaudily-coloured shrimp poses in an obvious place on a coral reef. Fish are attracted, and they allow the shrimp to clamber over them unmolested to clean off any parasites—and these are fish that would normally eat any shrimp they encounter.

Crabs and prawns are caught as food for

72

Above: A hermit crab, living inside the old shell of a common whelk. Hermit crabs have soft bodies and so need to find homes in which to shelter.

Man all over the world. Big prawns are farmed in Japan and freshwater crayfish are farmed in parts of Europe. The Carib Indians in the West Indies used to eat lots of land crabs; we know this because their middens (refuse heaps) contain quantities of crab shells. Possibly the smallest crustacean to be exploited commercially is a copepod that is caught in Norwegian fjords. It is used in trout farms to feed the fish during the winter.

Eating the whales' food

Many sea animals including the largest of all, the blue whale, feed mainly on krill, tiny shrimp-like animals which live in the oceans in vast quantities. Penguins and crab-eater seals are also krill-eaters. It now seems likely that we shall also be eating krill. Scientists are working out how to process krill to provide an important source of protein in our diet. Conservationists, however, are worried that if too many krill are fished the sea animals may go short of food, which could endanger some species. The danger is probably less, however, than if we continue to kill whales at the present rate.

Guide to crustaceans

Cephalocarida Tiny animals living in the sea bed, discovered in 1955.

Branchiopoda Water-fleas and other fresh-water crustaceans, and also some sea animals.

Mystacocarida Minute seashore animals, first found in 1943.

Copepoda Small water animals, many of them parasites.

Branchiura Fish-lice.

Ostracoda Seed shrimps.

Cirripedia Barnacles and allies, including some parasites.

Malacostraca All the larger crustaceans, including lobsters, crabs, shrimps, prawns, woodlice.

Spiders and Their Allies

Arachnids are an important group of arthropods—important because they help to keep the insect population of the world in check. The most familiar members of the class are the spiders. The others include scorpions, harvestmen, ticks and mites.

Arachnids have eight legs; this enables you to tell them immediately from insects, which have six. They do not have wings, and their bodies are in two parts, the abdomen and the cephalothorax, which is a combined head and thorax. They do not have antennae (feelers), but they have two other pairs of appendages. One is a pair of chelicerae, which are like either a pair of pincers or a pair of fangs. The animals use them for feeding. Behind the chelicerae is a pair of pedipalps, which are leg-like structures, but not used for walking. Arachnids have up to four pairs of eyes. Nearly all arachnids are carnivores, and some of them kill their prey by injecting poison.

Scorpions are the oldest group of arthropods, and they have been around for about 400 million years. They live in the warmer parts of the world, and there are about 800 known species. Scorpions range in size from 13 millimetres ($\frac{1}{2}$ inch) long to the giant *Pandinus* scorpion of Africa which measures 180 millimetres (7 inches). The scorpion's chelicerae are a formidable pair of pincers, looking a bit like those of a crab. Its abdomen is elongated to form a tail which has a sting in its tip. The scorpion can bring its tail over its body to sting its prey or an enemy. A few species produce poison which can kill a man. Scorpions produce living young, from a few months to a year after mating.

Pseudoscorpions look a bit like scorpions, but have no sting. They are tiny, the biggest being only about 8 millimetres (1/3 inch) long.

Below: A female wolf spider, with a cocoon containing her eggs, climbing up the stem of a plant.

Bottom: A scorpion, with its long tail curled round at the rear.

They are rarely seen, but you can find them in leaf mould in most parts of the world. They feed on smaller arthropods.

Whip scorpions have long tails that look like whips. They live in the tropics or semi-tropical areas and hunt only at night. They can spray an attacker with a defensive fluid which smells like vinegar and can burn the skin. From this habit comes their popular name of vinega-roons. They range in size from 2 millimetres (2/25 inch) to 65 millimetres (2$\frac{1}{2}$ inches).

Harvestmen, which look like very long-legged spiders, are also arachnids. They live in all parts of the world, and there are 3,200 species, ranging in size from tiny ones 1 millimetre long to tropical giants with legs 160 millimetres (6 inches) long. They are scaven-

gers as well as flesh-eaters. In North America they are often called daddy-long-legs, which in Britain is what people call the crane-fly.

There are 25,000 species of mites and ticks, and they live almost everywhere in the world, from polar ice to hot springs. Most of them are parasites on plants or animals, and are pests causing damage to crops and domestic animals. Most of the mites are very tiny indeed, but some ticks are as much as 30 millimetres (1$\frac{1}{8}$ inches) long.

Spiders are the most familiar of the arachnids, and there are at least 32,000 species. There are enormous populations of some species: a zoologist once calculated that 1 hectare (2$\frac{1}{2}$ acres) of grassy meadow in Britain contains well over 5,500,000 spiders. Spiders all produce silk, and the various species put this silk to different uses. The familiar web

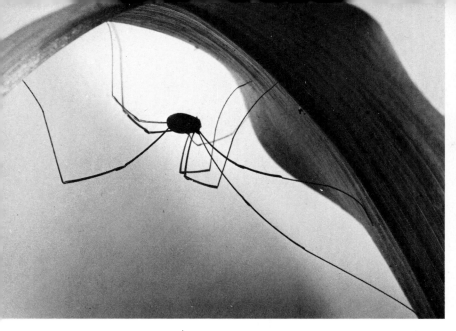

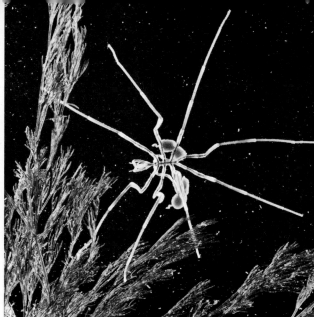

spiders construct large, radiating orb webs with this silk. Others make large sheets or tangles, or line their nests with silk. A spider leaves a safety-line of silk behind it as it moves about, which is why if you dislodge a spider it is found swinging in mid-air, clinging to its dragline. Spiders lay eggs, and their silk is used to make a protective cocoon for the eggs. Almost all spiders produce some venom, usually just strong enough to paralyse their normal prey. A few, particularly the black widows, produce poison powerful enough to affect Man; but even then few black widow victims die.

The main group of spiders include the large species called variously tarantulas, bird spiders or monkey spiders; the garden spiders which spin orb webs; the black widow and its relatives which make loose webs; and various kinds of hunting spiders. The jumping spiders hunt by sight and stalk their prey, while wolf spiders chase their victims over open ground. Trap-door spiders live in burrows, closed by silken doors beneath which they lurk.

Above left: A harvestman clinging to the underside of a leaf.

Above right: A male sea spider—which is not an arachnid but a close relative—carrying eggs.

Below: The horseshoe crab or king crab.

Guide to arachnids

Scorpiones Scorpions.
Pseudoscorpiones False scorpions.
Solifugae Sunspiders, also known as wind scorpions.
Palpigradi Tiny scorpion-like animals.
Uropygi Whip scorpions.
Amblypygi Tropical spider-like animals.
Araneae The spiders.
Ricinulei Heavy-bodied animals living in leaf-mould in Africa and America.
Opiliones Harvestmen.
Acarina Mites and ticks.

Arachnid relatives

Two groups of animals are similar to arachnids, but are placed in classes by themselves. The horseshoe crabs, also called king crabs, form the class Merostomata. The horseshoe crab has a horseshoe-shaped upper shell, and a long spike at the rear which the animal uses for pushing into the sea-floor mud where it lives, and also for righting itself if it is turned over. It has five pairs of walking legs and a pair of pincers. The female horseshoe crabs lay their eggs in the soft sand of the seashore.

Sea spiders form the class Pycnogonida, and they live in all the oceans. They look very like spiders, but some species have 12 legs. Most are small, but some living near the Poles have a leg-span of more than 600 millimetres (2 feet). There are 600 species.

Insects by the Million

Insects are the most numerous of all animals. More than 1,000,000 different species have been identified and classified, and entomologists, scientists who study insects, are discovering more all the time. As for the numbers of each species, they are beyond counting.

Adult insects have six legs, which is how you can tell them easily from spiders and other arachnids. Their bodies are in three sections, the head, the thorax and the abdomen. Like other arthropods, insects have an exoskeleton, a hard outer casing. Nearly all adult insects have wings. An insect's mouth is more complicated than a mouth as you know it, and is generally referred to as its mouthparts. Some insects have mouthparts adapted for chewing their food. Those of others, such as butterflies, are designed for sucking food in liquid form, such as the nectar of flowers. Insects which feed on the juice of plants and those which suck the blood of other animals have mouthparts with a long, piercing beak.

Most insects have compound eyes, with up to 30,000 tiny lenses, called facets. They also have senses which respond to scent, chemicals and touch, while some insects, especially the sound-producing ones such as grasshoppers, have organs which provide a sense of hearing.

There are three basic kinds of insects. Apterygote insects are the simplest, and include the familiar silverfish. They have no wings, and the young when they hatch from the egg look like smaller versions of the adults. The word apterygote means without wings. The other two groups of insects are the Pterygota—the winged ones, though some, such as fleas and lice, no longer have wings. The Pterygota are divided into two groups according to the way their young develop. This development is called metamorphosis because of the marked changes that occur at each stage.

Endopterygote insects go through a four-stage metamorphosis. They include beetles, bees, butterflies and flies. The insect begins its life as an egg. When it hatches from the egg it is a larva or grub, looking quite unlike the adult insect. The caterpillars of butterflies and moths and the maggots of houseflies are examples of larvae. In the larval stage the insect spends its time eating and growing. For the next stage of its life the insect makes itself a hard case, the chrysalis. Some species of insects spin a cocoon out of silk. Inside the case the larva becomes a pupa. In this pupal stage the body of the insect is completely broken down and remade. Eventually the case breaks open, and out comes an adult insect.

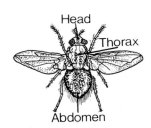

Above: Diagram showing the structure of a typical insect. This is a common bluebottle.

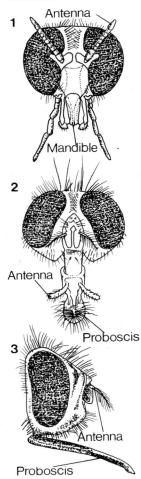

Above: Three types of mouthparts which are typical of insects: **1** shows a front view and the mandibles used for chewing, **2** another front view of an insect showing the proboscis used for sucking liquids, and **3** shows a side view of a fly with the piercing proboscis which is used for piercing the skin and sucking the blood of other animals.

Opposite: A head-on view of a common field grasshopper.

The exopterygote insects become adult in an incomplete metamorphosis, in three stages. There is no pupal stage in their development. The insect hatches out of the egg as a nymph, an immature form without wings but otherwise similar to the adult. The nymph moults as it grows, the stage between each moult being known as an instar. Finally, it becomes an imago, or adult, with fully-formed wings. Dragonflies, crickets, earwigs and true bugs develop in this way.

Guide to insects

APTERYGOTA (*wingless insects*)
Thysanura Silverfish.
Diplura Two-pronged bristletails.
Protura Tiny, sightless insects that live in the soil.
Collembola Springtails.

EXOPTERYGOTA (*winged insects with three-stage metamorphosis*)
Ephemeroptera Mayflies.
Odonata Dragonflies and damselflies.
Plecoptera Stoneflies.
Grylloblattodea Small insects that live in the soil.
Orthoptera Crickets, grasshoppers and locusts.
Phasmida Stick and leaf insects.
Dermaptera Earwigs.
Embioptera Small, tropical insects that live in woven silk tunnels.
Dictyoptera Cockroaches.
Isoptera Termites.
Zoraptera Rare, tiny, agile, tropical insects, living in termite nests.
Psocoptera Booklice.
Mallophaga Biting lice.
Anoplura Sucking lice.
Hemiptera The true bugs.
Thysanoptera Thrips.

ENDOPTERYGOTA (*winged insects with complete metamorphosis*)
Neuroptera Lacewing, snake and alder flies.
Mecoptera Scorpionflies.
Lepidoptera Butterflies and moths.
Trichoptera Caddis flies.
Diptera True flies.
Siphonaptera Fleas.
Hymenoptera Bees, wasps, ants, sawflies.
Coleoptera Beetles.
Strepsiptera Parasites with twisted wings.

Insects: The Beetles

One-third of all the different species of insects consists of beetles. Although there are large numbers of each of these 330,000 species, you do not see them as often as other kinds of insects, mainly because beetles tend to be more retiring in their habits. Many of them live down in the soil or in rotting animal and vegetable material such as leaf mould, dung, carrion, decaying wood and fungi. Others live in water, and most of the species fly, so they can be found in earth, air and water. They eat almost anything, though each species has a limited, specialised diet.

The beetles' scientific name is Coleoptera, which means 'sheathed wings', and it perfectly describes these fascinating animals. In a beetle, the front pair of the standard two pairs of wings, which most insects have, has been transformed into tough, horny cases. These cases close to protect the other, delicate, pair of wings with which the insect flies. Most beetles also have tough body armour on the underside as well. This protective outer casing helps to guard beetles against birds and other predators. Most beetles are dull coloured, but some species have very brightly-hued wing cases.

The mouthparts of beetles are constructed for chewing solid food, and so are those of

Right: One of the largest species of beetles, the goliath beetle of central Africa. Specimens have been found up to 175 millimetres (7 inches) long. A seven-spot ladybird is shown with it for comparison.

Below: The sexes of many beetles differ greatly in size and shape, a condition called dimorphism. The male of these two stag beetles is the larger one with the huge jaws.

Question of size

Beetles come in all sizes (see left). The biggest are the goliath beetle, the rhinoceros beetle and the long-horned beetle, all tropical species. Specimens of these three kinds have been found up to 175 millimetres (7 inches) long. By contrast, the smallest beetle is another semi-tropical species, *Nanosella fungi*, which is only 0.25 millimetres (1/100 inch) long.

Guide to beetles

Classifying beetles when there are so many different kinds is a very complicated business, and there are a great many different families. The main division of the order Coleoptera is into the following four suborders:

Archostemata A small, wood eating group of about 20 species.

Myxophaga Another small group found in wet places.

Adephaga These are the most primitive of all beetles. They are active predators. There are eight families, of which one, the Carabidae, contains 25,000 species.

Polyphaga This suborder contains the majority of beetle species, in many different families grouped into 20 superfamilies.

Above: *Dytiscus,* the great diving beetle, is a fierce predator. It will attack any prey if it is not too large.

their larvae. Most beetles have eyes, but some species living in caves and elsewhere underground do not. Beetles develop by means of complete metamorphosis (see pages 76-77).

Many beetles are pests because they eat our crops. The wireworms that eat roots of cereals and other plants are not worms, but the larvae of click beetles. These beetles get their name because they make a sharp clicking noise when they jump.

There is a whole group of beetles whose larvae are scavengers, eating dead or decaying animal or vegetable matter. In the wild they do a good service, helping to dispose of dead material, but unfortunately they attack such things as leather, fur, wool, grain, bacon and cheese. Another group have larvae which spend their time boring into dead wood and eating it. In this way they help to dispose of rotten wood and dead trees in the forest—but in our homes they eat furniture and the timbers of which buildings are made. You probably know them as woodworms and death-watch beetles.

Another group of beetles preys on insect pests such as aphids. Ladybirds or ladybugs are the best-known of this kind. Others eat the leaves and flowers of plants. Leaf-miners are tiny larvae which are so small that they are able to live between the upper and lower surfaces of a leaf, tunnelling through the soft green pulp between them.

Among the more extraordinary beetles are the bombadier, which when attacked ejects a little jet of caustic liquid with a pop; the short-circuit beetle, also called the lead borer, which eats holes through the lead casing of telephone cables in the United States; and stag beetles, which have huge mandibles looking like miniature antlers.

Of the various water-beetles the best-known is *Dytiscus,* the great diving beetle. It is about 40 millimetres (1½ inches) long, and the female dives into ponds to lay her eggs in the stems of water plants. These beetles spend most of their lives in the water or close to it. Whirligig beetles spin round on water.

Many beetles can make chirping sounds, a process known as stridulating. They do this by rubbing one part of the body, called the scraper, against another called the file. They apparently do this to call to each other at mating time.

79

Social Insects

Most insects are solitary animals. They live on their own, meeting only to mate, and few of them have anything to do with their offspring once they have laid their eggs. However, there are four kinds of insects that have highly organised community lives. Only Man and some other mammals have anything similar. We call these insects the social insects. Three kinds of insects are related: bees, wasps and ants all belong to the order Hymenoptera; the other kind, the termites, are in the order Isoptera, though since their appearance and life-style are similar to those of the ants they are often wrongly called white ants.

The most complicated bee society is that of the honeybees. Each hive contains three different kinds of bees. There is one queen, a large insect whose sole task is to lay eggs. Then there are several hundred drones, males which do no work. Finally there are the worker bees, female bees which are not capable of laying eggs. There may be up to 80,000 of them. A worker bee lives for about six weeks, seldom more. For the first three weeks after the bee becomes an adult she is engaged on 'housework', duties around the hive. These include cleaning out, building and repairing honey-combs—rows of cells made of wax—looking after the larvae and tending the queen. For the rest of her life she goes foraging, flying in search of nectar and pollen and bringing them back to the hive to be made into honey. The drones stay around the hive, eating food supplied by the workers. When new queens are born and fly, the drones fly with them and mate with them. Then the drones die.

Ants have an even more highly-organised way of life than honeybees. Like the bees, ants are of three kinds—queen, males and workers (who are female). Many species of ants have more than one size of worker, with different duties. The larger sizes act as soldiers or sentries, protecting the nest. A typical ants' nest is a maze of corridors and chambers, each room with its own purpose. Some house newly-laid eggs, others the young larvae and the pupae. Some rooms are set aside as store-chambers or rubbish dumps, and a few species have gardens, where they grow a kind of fungus which they use for food. Ants' nests vary greatly: some are hidden underground, others are built up to form the familiar anthills. A favourite place for a nest is under a flat rock. On a sunny day, when the rock is hot, the ants move the eggs up close under it to keep them warm. If you lift the stone, within minutes the ants will have taken the eggs down below for warmth and safety. Ants

live longer than honeybees, some species for as much as six years, with the queens living to 15. The size of an ant colony varies from fewer than 100 to more than 100,000.

Marvellous though the community life of ants is, that of the termites is even more remarkable. Nearly all termites live in tropical or semi-tropical lands. Like the ants, termites have a queen, whose only job is to lay eggs, but they also have a king, a male termite who is much smaller than the queen. He mates with her when the colony is founded, and then lives with her. Both queen and king live for a long time, over 50 years in some species.

The remaining termites are workers and soldiers, whose ranks include both males and females. Generally the smaller workers stay in the nest and do the housework, while the larger ones go out foraging or tunnelling in search of food. A termite nest is the finest example of animal architecture. It is often built to a considerable height above ground, and has internal shafts which provide its own air conditioning. Up to 1,000,000 termites may live in one nest.

Most species of termites live broadly similar

A soldier ant from the tropical rain forests of Peru. This huge ant is 25 millimetres (1 inch) long, and has greatly enlarged mandibles.

lives, but those of ants and bees vary greatly. Army ants have no fixed abode, but spend their time travelling, making a temporary camp for a week or two while they eat everything within foraging distance. A great many species of bees and their close relatives wasps also have a community life, but their colonies are much smaller, and many of them last only during the summer season. At the end of that time new queens mate and go into hibernation (see pages 98-99), and the rest of the colony dies. Solitary bees and wasps make nests, provide a store of food for the young, and after laying their eggs take no further interest in them, though a few take extra food to the larvae.

Above: Gathering food, a bumblebee seeks nectar and pollen from a French marigold.

Above right: Paper wasps on their nest, which is built on a prickly pear cactus in Spain. Wasps were the first animals to make true paper.

The three kinds of honeybees: left, queen; centre, worker; right, drone.

Guide to the bees, wasps and ants

The order Hymenoptera, to which bees, wasps and ants belong, contains more than 100,000 species. It is divided into two suborders, each of which contains several superfamilies.

SUBORDER SYMPHYTA

Insects in this group do not have a narrow waist between the thorax and the abdomen. Its members are either wood-wasps, which bore into wood to lay their eggs, or sawflies, which saw into plant tissues for that purpose. There are six superfamilies.

SUBORDER APOCRITA

Nine out of ten species belong to this group. There are 14 superfamilies. Most of the members of seven of these superfamilies are parasites, some of them very small. The other superfamilies are:

Ichneumonoidea Ichneumon flies, some of them quite large, whose larvae are parasitic on other insects.

Bethyloidea Cuckoo wasps, which make use of other wasps' nests.

Scolioidea Ants and parasitic wasps.

Pompiloidea Spider-eating wasps.

Vespoidea Social wasps.

Sphecoidea Solitary wasps.

Apoidea Bees.

Flies of All Kinds

A very large group of insects goes by the general name of flies. The most familiar are the houseflies and bluebottles that are pests in the home, and they belong to the order Diptera, generally called the true flies. The name of the order means 'two-winged', and that is just what the true flies are. Only one pair of wings is functional. The rear pair has changed into two club-like growths, called halteres.

More than 85,000 species of true flies are known, and they have a variety of life-styles. Most of them fly by day, and a great many feed on the nectar and pollen of flowers. Hover-flies are a good example. Others, including the housefly, dung-flies and blow-flies, like to feed on carrion. They also like to settle on exposed food, particularly meat, and they carry germs from one meal to another on their legs and in their saliva. For this reason they are major carriers of disease. These flies do not bite, but a great many do, including mosquitoes, horseflies, botflies and biting midges. Mosquitoes are also great carriers of disease, because they suck blood from their victims. Like houseflies, mosquitoes carry the germs of diseases in their saliva. Among the illnesses they transmit are malaria and yellow fever. Only some mosquitoes carry disease in this way, and only the females do so; the males do not suck blood. Members of another fly family, tsetse flies, are even more deadly as carriers of diseases. They live in parts of tropical Africa.

Of the other insects with 'fly' as part of their names, one of the most interesting groups is that of the dragonflies and damselflies. They are elegant, long-bodied animals with long,

Above: A lacewing is a kind of flying insect, but it does not have the word 'fly' in its name. Many of them are green like the leaves on which they lay their eggs. Lacewings are in the order Neuroptera.

Left: The larva of a caddis fly emerging from its case under the water.

Left: A dragonfly can fly forwards and backwards, and also hover. Its filmy wings enable it to travel very quickly.

lacy wings. Damselflies have two pairs of almost identical wings, while dragonflies have broader hind wings than front wings.

There are 5,000 species of dragonflies. They feed as they fly, eating other insects such as mosquitoes. Dragonflies fly very fast, and some have been known to reach 100 kph (60 mph). They mate in the air, and the female lays her eggs in water or on the stalks of water plants. The nymphs live in the water for up to five years before turning into adult insects.

Most dragonflies are only a few centimetres long across the wings, though some tropical species have a wingspan of up to 190 millimetres ($7\frac{1}{2}$ inches). Fossil dragonflies have been found with a wingspan of more than 600 millimetres (2 feet), and these were apparently among the first flying insects known.

You will find two other groups of flies near water: stoneflies and caddis flies. Stoneflies live in the water as nymphs, and spend their adult lives crawling about among the stones on river banks. They fly very little, preferring to scuttle about the ground. They look a bit like grasshoppers, and their dull colours make them hard to detect. Caddis flies are equally dull in colour, and because they fly mostly at dusk they are equally easy to overlook. Their larvae are much better known. They are soft-bodied creatures, living in ponds and streams, and to protect themselves they make cases. These cases are made with silk thread, which is used to bind together materials from the water around them—grains of sand, bits of rock, small shells, or pieces of plants.

Scorpionflies spend most of their lives, both as larvae and as adults, in or on the soil, and some of them have only small wings and hop somewhat like grasshoppers. There are only about 400 species.

Above: A blowfly feeding on decaying matter. These insects carry disease.

Below: A crane-fly is one of the true flies, related to houseflies, gnats and mosquitoes. It eats plant matter, and many species of crane-flies are serious pests in gardens.

One day of life

The mayflies get their Latin name, Ephemeroptera, from two Greek words meaning 'living a day' and 'wing'—and that is just what happens to the adults. They emerge from life as nymphs in ponds or streams to fly for perhaps as little as a few hours. In that time they mate. You can see the males in a cloud, dancing over the water. Females fly into the cloud, are mated, lay their eggs, and die. Adult mayflies do not eat, and the longest any species lives is a week. However, as nymphs they have an underwater life of up to three years.

Guide to flies

There are five orders of insects that have the word 'fly' in their names, though of course they are not the only flying insects.
Ephemeroptera The mayflies.
Odonata Dragonflies and damsel-flies.
Plecoptera The stoneflies.
Mecoptera Scorpionflies.
Diptera The true flies, including houseflies, blowflies, fruit flies; mosquitoes; midges; crane-flies; horseflies.
Trichoptera The caddis flies.

Insects: Butterflies and Moths

Probably the most beautiful of all the insects are the butterflies and moths. There are more than 100,000 different species, and together they make up the insect order Lepidoptera. This name comes from two Greek words meaning scale-wing, and if you can look closely at a butterfly's wing you will see that it is covered with tiny scales. The often vivid colouring of butterflies and moths is built up by these scales, like a mosaic picture.

One difficulty is to decide when an insect is a butterfly and when it is a moth. As a group the insects have many features in common, and many moths are similar in type to butterflies. However, as a rough and ready rule you can take it that most butterflies fly by day, and moths at night or in twilight. When they are resting, most butterflies sit with their wings folded together, upright, while moths rest with their wings outstretched. A butterfly's antennae end in small, bare knobs, while nearly all moths have antennae with pointed tips. Most moths have plump bodies, while the bodies of nearly all butterflies are slender. Finally, most of the brightly-coloured members of the order are butterflies, and moths flying at night are in more sombre colours which make them harder to pick out.

All members of the Lepidoptera have a long sucking tube, called a proboscis. This is normally curled up, but the insect can extend it, like a party 'squeaker', when it wants to suck nectar from a flower. The insects have two pairs of active wings, giving them a big area.

Although the adult stage of a butterfly's or moth's life is the most beautiful, it is the early part of the life-cycle that is the most important from our point of view. All these insects produce caterpillars, which feed on plants.

Like the adult insect, a caterpillar has six legs, attached to the thorax. It also bears five pairs of extra limbs, called prolegs, on its abdomen. These prolegs end in hooks which enable the caterpillar to cling firmly to the plant on which it is feeding.

Caterpillars produce silk, which they use for a variety of needs. The main use of the silk is to spin a cocoon when the animal comes to pupate. Nearly all species do this.

Although the larvae of butterflies and moths are destructive, and do great damage to crops, the adults perform a valuable service in pollinating flowers. Despite their fragile appearance, these insects are strong fliers, and can cover amazing distances. Several species migrate from one place to another according to the season, one of the most remarkable migrants being the monarch butterfly of North America (see pages 94-95).

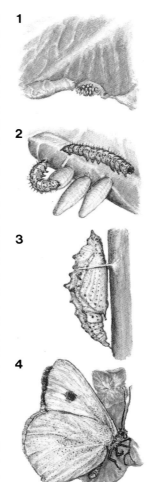

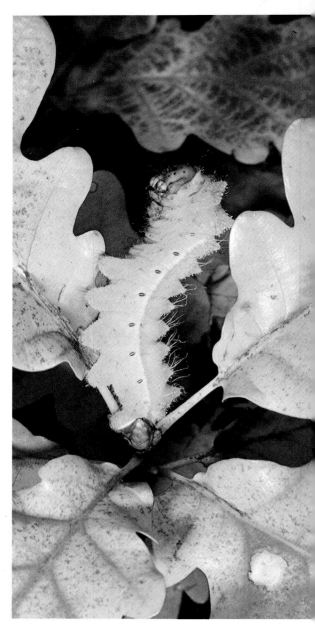

The four stages of metamorphosis: **1**, eggs on a leaf; **2**, caterpillar emerging from the egg; **3**, the chrysalis; **4**, the adult butterfly.

Above: A peacock butterfly feeding on a thistle.

Above left: The caterpillar of an Assam silk moth feeding on an oak tree.

Left: A small elephant hawk moth at rest on a flower.

Dodging the enemy

Moths which fly by night escape the attentions of hungry birds—but they fall victim to the only flying mammals, the bats. Bats have developed a sonar (sound-detection) system by means of which they can locate moths, swoop on them and catch them.

The moths have developed various methods of escape. Many species have evolved sensitive hearing, so they can pick up the signals of the bats' echo-sounding device, and take evasive action. Some fold their wings and drop like stones, while others loop, spin and turn to get out of the way.

Stealing the silk

The silk we use for clothing comes from the cocoons spun by the caterpillars of the silkworm moth, or *Bombyx mori* to give it its scientific name. These caterpillars produce the strongest and most plentiful silk. They feed on mulberry trees.

Silk was discovered about 2700 B.C. in China—according to legend, by the Empress Hsi Ling-shi. The Chinese guarded the secret for many years, but in the AD 550s the Roman Emperor Justinian sent two monks to China to steal it. They smuggled out a supply of silkworm eggs in bamboo canes, and so brought the silk industry to Europe.

Insects: Hoppers and Crawlers

There are several other important groups of insects besides those species described in the past ten pages. Among the most interesting is the order Orthoptera, whose 10,000 species make up the crickets, mole-crickets, grass-hoppers, locusts and katydids. Nearly all of them can fly, but they usually move by jump-ing, and they have long, powerful hind legs which enable them to do this. 'Hoppers' are perhaps best known by their loud, chirping 'songs'. These sounds are made only by the males, except in a few species, and they produce them by rasping a scraper against a file. The file is on the forewings, and the scraper may be either on the legs or the wings. The song is used in courtship. They hear with 'ears' on the body or on the front legs.

Some members of the Orthoptera eat other insects, but most are vegetarian. Seven species of short-horned grasshoppers (those with short antennae) are the notorious locusts, which have been devastating crops for thousands of years. Normally the locust is an inoffensive green grasshopper, existing in small numbers. Every so often there is a population explosion. When the grasshoppers become overcrowded they change colour to a dark reddish brown, and their character changes too. Eventually enormous swarms of these insects take off and fly across the land. They arrive in a cloud so dense that it makes the sky dark, and settle on every plant. When they leave, there is nothing left—they have eaten every leaf and every blade of grass.

The word 'bug' is misused for any insect, but the name belongs to the 50,000 species of one order, the Hemiptera, the true bugs. They are mostly flat-backed animals, with mouth-parts which are adapted for piercing and sucking up liquids. They live mostly on the juices of plants, and aphids—the familiar greenfly—are the best-known species. Some are very small, but a few species of bugs are more than 100 millimetres (4 inches) long.

Another familiar group contains the ear-wigs, who form an order on their own, the Dermaptera. Their popular name comes partly from their formidable pincers, which look a bit like a tool formerly used for ear-piercing, and also because they hide in small cracks and nooks and were believed to conceal them-selves in people's ears. They eat mostly at night, browsing on vegetable matter and also scavenging.

Three groups of insects are considered real pests by Man. They are the lice, fleas and cockroaches. Lice and fleas are parasites living on larger animals.

There are two kinds of lice. The chewing or

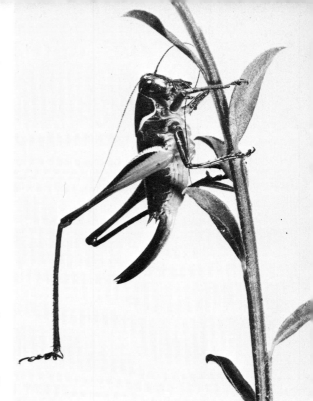

Right: A female bog cricket cleaning her antennae.

Below: A dense swarm of locusts over an airport in Somalia. The plane had been spraying the swarm, but there were so many insects that they choked the engine and it would not start.

Bottom: A female praying mantis. Her egg case is on the twig behind her.

biting lice belong to the order Mallophaga, and they are all very small, the biggest being only about 10 millimetres (3/8 inch) long. They have flat bodies and no wings. There are more than 2,600 species, and each is adapted to life as a parasite on a particular bird or mammal. If their host dies they die too, unless they can transfer to another animal of the same species. The sucking lice, in the order Anoplura, live on mammals and suck their blood. There are at least 300 species, of which two can infest Man.

The so-called booklice are not lice at all, nor are they parasites. They are members of the order Psocoptera whose species are adapted to eating fragments of animal or vegetable matter. The booklice get their name because they feed on the paste and leather used in binding books.

Like the true lice, fleas are wingless parasites, sucking blood from mammals and birds. They also have flat bodies, but in contrast to lice their bodies are flattened vertically. They have powerful legs, and the common flea can jump 130 times its own height. Some stick only to one kind of host, but others migrate from one species to another. They transmit disease when they drink the blood of an infected animal then bite a healthy beast. The worst disease they transmit is bubonic plague—the Black Death of the history books.

Cockroaches are disliked because some species invade people's homes, especially kitchens and larders, in search of food, but most of them live out of doors. They form the order Dictyoptera, and also in this order are the praying mantises, grasshopper-like insects which hold their huge forelegs in an apparently praying position. Mantises are carnivores, and some of the big ones even attack small frogs and birds.

Above: A common field grasshopper feeding on a spray of heather.

Mobile sticks and leaves

The tropical forests of Asia are full of some of the most bizarre insects, which look remarkably like portions of plants. They are the stick and leaf insects, which together form the order Phasmida.

They are very slow-moving creatures, and they rely for their safety on the fact that they can blend into their surroundings. Leaf insects place themselves on twigs in just the position that a real leaf would be, while stick insects look remarkably like the twigs themselves.

Right: Which is the twig? A stick insect takes up its camouflage position.

87

Microscopic Animals

In 1674 a linen-draper in Amsterdam wrote to the Royal Society in London to report that he had seen 'animalcules a thousand times smaller than the eye of a louse' in a drop of stagnant water. In this way people first learned about the existence of protozoans, the smallest of all living animals. The draper was Antony van Leeuwenhoek, an amateur scientist, and he was one of the pioneers in making and using microscopes.

The phylum Protozoa consists of animals which have a body made of just one cell. As is described on pages 14-15, cells are the basic units of life, and every plant and animal is made up of them—usually many millions of cells. The largest protozoans are about 5 millimetres (1/5 inch) long, but most of these animals are so tiny that you need a microscope to see them, and a powerful one to study their structure. Some scientists today include them with the simplest forms of plant life in a third kingdom, the Protista (see pages 12-13).

Nearly all protozoans live in water—both fresh and salt—but some also live in the soil. Some are parasites on or in other, larger animals. They move about in various ways, and zoologists classify them partly according to these ways. Some, such as the amoeba, move by pushing out a part of the cell wall to form a pseudopodium—a false foot. The rest of the animal then flows into this pseudopodium. Another group moves by means of flagella (beaters). A flagellum is a long thread-like organ which projects from the wall of the cell. It undulates or ripples with a wave-like motion, and so propels the cell along. Many protozoans have cilia, which are similar to flagella but much finer.

Protozoans also have varied ways of reproducing. Some divide to form two new cells, each with all the features of the original cell, and approximately equal in size. Others produce projections from the cell wall, called buds, which eventually break away and form new protozoans.

Like other animals, protozoans need food, and they get this from the water or other surroundings in which they live. The food is usually in the form of minute particles or even substances such as proteins dissolved in water. Amoebas and similar protozoans absorb their food by wrapping a pseudopodium around it and drawing it into the cell body. Others draw liquid into deep cones in the cell wall, and then absorb it. Protozoans with cilia use them to waft food particles into the cell's 'mouth', a funnel-shaped opening. Some protozoans contain the substance chlorophyll, which makes plants green, and use it—like plants—to make their own food (see pages 112-113).

Although they are so small, protozoans are of great importance in nature. Countless millions of them form part of the plankton, the mass of tiny animal and plant life found in the surface waters of the oceans. Others have minute limy shells, and when they die these shells sink to the bottom of the water, where in due course they form limestone. Many of the parasitic protozoans cause diseases, such as malaria and sleeping sickness.

Right: An amoeba putting out many 'arms'.

Far right: The cellulose cell wall encircles these dinoflagellates.

Below left: A colony of the protozoan *Volvox* producing daughter colonies.

Below: *Paramecium* in the middle of red-stained yeast cells, on which it is feeding.

Below right: These star-like animals are radiolarians, which live in the surface waters of the great oceans.

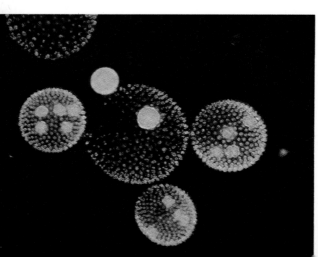

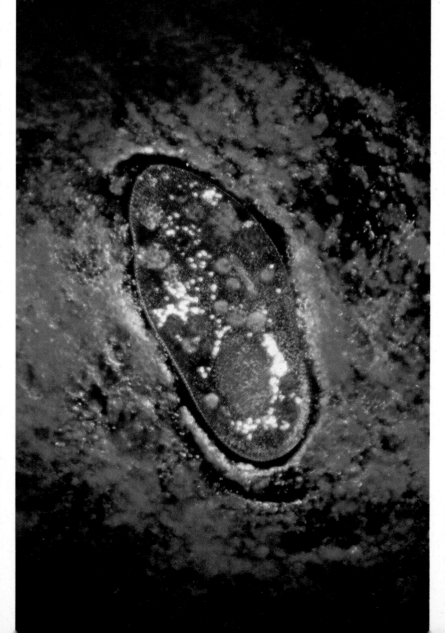

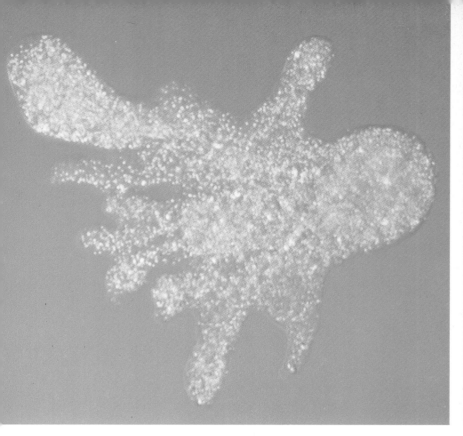

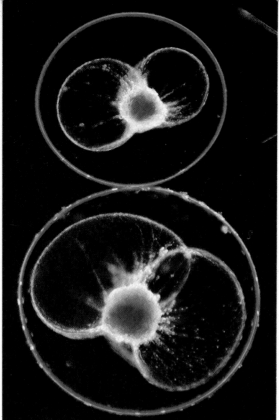

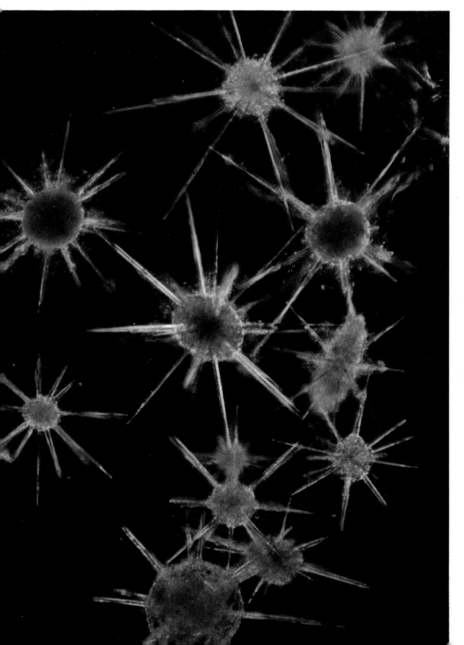

Guide to protozoans

Scientists do not agree on the ways that protozoans should be classified, but the following is the most usual method:

Mastigophora Flagellates; some are like plants and produce their own food. Many others are parasites.

Rhizopoda -The amoebas and also the foraminifers, protozoans which have chalky shells.

Actinopoda Heliozoans—'sun animals' which appear to have rays like the Sun; and radiolarians, which have beautiful shells.

Sporozoa Parasitic protozoans, of which some cause malaria.

Cnidosporidia Small parasites.

Ciliata Protozoans with cilia.

Sponges

Sponges are very primitive animals made up of many cells, but these cells lead a fairly independent life. The sponges have skeletons, made up of fibres or tiny needles, and it is these skeletons which people use for cleaning.

There are about 10,000 species of sponges, most of them living in shallow sea water. They form a phylum on their own, the Porifera.

Instinct and Learning

When you look at a beautifully-woven spider's web or see swallows setting off on their long migration flights, you may well ask, 'How do the animals know how and when to do these things?' You may think that termites, which build vast nests with air-conditioning systems, are very clever, and wonder how they can be when their brains are so small compared with that of Man.

Biologists have been trying to find the answers to these and other questions for a long time. These answers—as far as we now know them—are fascinating, and far from the obvious ones, for a great deal of the behaviour of animals is the result of what we call instinct. The animal is pre-programmed, like a computer, to do certain things. During its lifetime the programme may be modified or added to, by the process we understand as learning.

The foremost point to be understood is that animals do not behave and think in the same way that we do. For one thing, they have very

A bird's nest, such as the one built by this rufous fantail (an Australian flycatcher), is carefully woven, but the bird makes it by instinct—no other bird shows it how to do this.

different senses. Humans rely basically on sight, hearing, touch and smell to tell them what is happening in the world around them. Many animals possess some or all of these senses, but they work in different ways from ours. For example, some animals are colour-blind, and see everything as shades of grey. Golden hamsters do this. In spite of the old saying 'Like a red rag to a bull', bulls cannot in fact see the colour red—whereas penguins and gulls can. Bees cannot see red, but they can see ultra-violet at the other end of the spectrum, which is just invisible to us. Although many animals cannot see nearly as well as we do, some see much better. Birds of prey hovering high in the sky can detect a small animal on the ground at a distance where people would need to use binoculars.

Probably the sense which is most different from ours is that of smell. Most animals can pick up a whole range of odours that pass us by undetected. Some of these smells, which lure males to females at mating time or help to keep colonies of insects such as bees together, are called pheromones.

In addition, many animals, especially birds

and some insects, can apparently detect the Earth's magnetism, and react to it in much the same way as the needle of a compass does.

Instinct provides animals with a great many abilities. For example, a swallow turns to and builds a nest in the spring without having either built one before or been shown how to do so. The nest looks like any other swallow's nest. If it builds a second nest it probably does it better, because it has learned by experience. In this way learning reinforces instinct.

A very special example of instinct and learning working together was found by the Austrian naturalist Konrad Lorenz. He discovered that young ducklings become attached to the first thing—particularly a living thing—they see. They then follow it around and treat it as they would their mother. This is called imprinting, because the image of the mother or mother-substitute is imprinted on the ducklings' brains. The ducklings learn which is their mother-figure: instinct then tells them to follow it about.

What triggers off instinctive actions by animals? We know that a lot of the brain's work is controlled by tiny chemical reactions. In the same way, the instinctive actions controlled by an animal's brain can be triggered off by chemical reactions. This is how pheromones work. A smell is a chemical substance, and it can trigger off certain responses in an animal that detects it. In the same way, scientists think, other outside influences act on animals to trigger off responses: longer days induce birds to nest, shorter ones make those which are going to migrate prepare to set off.

Many of the higher animals can and do learn a great deal. If you keep a pet you already

Maternal instinct and the feeding instinct cross the barrier of species: a domestic cat suckles an orphaned squirrel.

The chimpanzee has learned how to feed himself with a spoon, but as a young animal he would suckle by instinct.

know that animals quickly become creatures of habit. Dogs, for example, expect to go for walks at certain times of day. In the days when delivery men made regular rounds with horses and carts, they found that the horses knew when to stop and start as well as they did. The animals have not only learned, but also remembered. Nobody teaches a dog to go and bury a bone: it does this entirely by instinct, as a store for lean times.

Chimps are tops for intelligence

Chimpanzees are the most intelligent animals. A scientist at Georgia State University reported in 1974 that a colleague had trained a female chimp to 'talk' a simple language by using a computer keyboard.

Other species of apes and monkeys have also learned to perform complicated tasks: a railwayman in South Africa is said to have trained a baboon to help him operate the signals, back in the 1880s, and more recently an Australian farmer claimed to have taught a rhesus monkey to drive a tractor.

Colours For Safety

Animals are engaged in a constant struggle to survive. To live, an animal must prey upon other animals or plants. In turn it must avoid being killed by other predatory animals. Camouflage—colour or shape providing a disguise—meets both these needs. It helps an animal to hide from its enemy, and it helps a predator to stalk its prey without being seen.

Animal camouflage takes many forms. Some animals, such as the African lion and the desert-dwelling fennec fox, blend into the background just by being the same colour as their surroundings. The polar bear and the Arctic fox are two other examples. Their white coats virtually disappear against their snowy habitats.

Countershading is another simple form of camouflage. Many animals are darker on top than they are below. Viewed from above, the animal blends into shadow; viewed from below, as a fish might be, the animal merges into the light shining behind it.

Irregular markings or mottlings such as spots and stripes also act as camouflage. They help to break up an animal's outline. Giraffes, tigers and zebras are hard to distinguish against long grass and foliage. Some animals can change their colour to match their surroundings. Sometimes the change is seasonal. The Arctic fox and Arctic hare have brown coats in the summer. As winter approaches they grow white fur.

The most remarkable colour changers are the animals that alter rapidly to match changes in their background. The chameleon is probably the best known, but it is not as expert as many other animals. The tiny Aesop prawn has an extraordinary range of colour tints—green to match seaweed, bluish-green, brown, red or violet to match coloured algae, and transparent blue at night. This prawn can change its colour completely within hours, sometimes in minutes. Various other animals—fish, frogs, shrimps, squid—also change colour. Plaice have even been known to adopt checked patterns to match checked linoleum beneath them, but the cuttlefish is the champion quick-change artist. It produces an entire range of colours from black to yellow, and can change from one to another within just a second.

Despite the variations in time, the mechanism by which animals change colour is much the same. The alterations are produced by the action of nerves and hormones on pigmented cells which lie just below the surface of the animal's skin. Patterns can be produced because each cell is separately controlled.

By contrast there are some animals, particu-

Above: This drone fly, a kind of hoverfly, mimics a bee—but it has no sting and is harmless.

Right: Flat-topped winkles, three yellow and six olive-green, 'hiding' on the frond of a rockweed—which has similar-shaped bulges called conceptacles on it.

Below: A buff-tip moth, looking exactly like a broken twig, rests on a similar-looking branch.

larly insects, that survive because they are brightly coloured or patterned. Instead of blending into their surroundings they stand out quite distinctly. Usually these are creatures that have stings or are poisonous. Their colours, which are known as warning colours, advertise the fact. Wasps and bees advertise their sting with bold black and yellow colours, and the monarch butterfly, which is distasteful to birds, is vividly patterned. Birds and other animals soon learn to recognise warning colours and to avoid the animals on which they see these hues.

Some quite harmless species also have warning colours or markings. The hoverfly is black and yellow, too, and looks in shape very much like a wasp. The viceroy butterfly resembles a monarch butterfly. Both these species can be eaten by birds, but they survive because of their colouring. This type of camouflage, in which one harmless species looks like another harmful animal, is known as mimicry. It occurs mainly among insects.

Indeed, insects display some of the most elaborate forms of camouflage. There are many insects that look remarkably like other objects, particularly leaves, sticks or twigs. The stick insect is one well-known example. The dead-leaf butterfly, when still, looks just like an old leaf, and when it is flying, its swirling motion makes it resemble a falling leaf. Some insects, such as the rose-leaf mantis, look exactly like flowers. There are even various moths, and a species of crab spider, that look just like bird droppings.

Some animals make their own disguises. One species of spider crab is able to 'dress itself up'. It places strips of seaweed, shells, and gravel all over itself until it is barely recognisable. If it moves into new surroundings it takes off the original camouflage and replaces it with a new one. Another crab, the sponge crab, camouflages itself with a piece of sponge, which it holds in place on top of its shell with a pair of specialised claws.

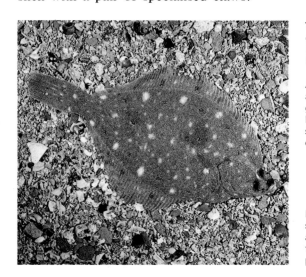

Top: A mountain grasshopper blends into its background of dead leaves and twigs.

Above: All is revealed: another mountain grasshopper raises its leathery forewing to show the bright colours of its abdomen.

Left: The spotted upper skin of a plaice makes it almost invisible when the fish is resting on a stony part of the sea bed.

Inky decoy

Most forms of camouflage depend on the animal keeping still—an object that remains motionless is far harder to detect than one that moves. An exception is the camouflage used by the squid.

This marine animal, when threatened with danger, squirts out a small cloud of dark ink, roughly its own size and shape. The inky decoy hangs in the water while the squid makes its getaway.

Migration

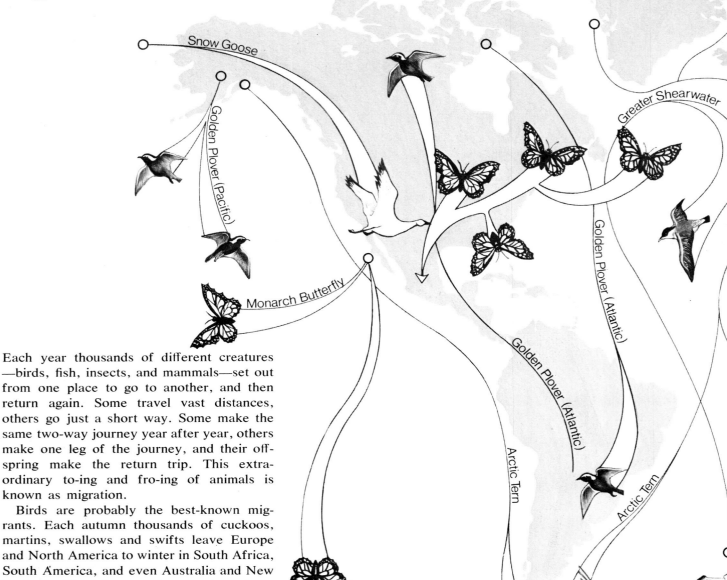

Snow Goose

Golden Plover (Pacific)

Monarch Butterfly

Greater Shearwater

Golden Plover (Atlantic)

Golden Plover (Atlantic)

Arctic Tern

Arctic Tern

Wandering Albatross

Each year thousands of different creatures —birds, fish, insects, and mammals—set out from one place to go to another, and then return again. Some travel vast distances, others go just a short way. Some make the same two-way journey year after year, others make one leg of the journey, and their off-spring make the return trip. This extra-ordinary to-ing and fro-ing of animals is known as migration.

Birds are probably the best-known mig-rants. Each autumn thousands of cuckoos, martins, swallows and swifts leave Europe and North America to winter in South Africa, South America, and even Australia and New Zealand. All of them are insect-eaters whose food disappears in the winter, forcing them to find supplies in warmer countries. In the spring they return again to breed.

It is the Arctic species of birds that fly the greatest distances. The Arctic tern is the most widely-travelled of all animals. Each year it migrates some 17,000 kilometres (11,000 miles) from the Arctic circle, where it breeds, to the Antarctic, where it spends the summer, and then makes as long a journey back.

Some land-living mammals, particularly those foraging in severe climates, also make seasonal migrations. The North American caribou spend the summer ranging over the Arctic tundra (plains). During the autumn huge herds of caribou move southwards into the forests. In about February they return again to the northern plains. Like birds, caribou follow the same routes year after year. Some mountain goats and sheep also migrate. They spend the summer on the upper slopes, and move down into more sheltered areas for the winter.

There are many migrating insects. One of these is the monarch butterfly. In summer monarchs are found throughout the United States and southern Canada. In the autumn vast swarms fly south, more than 2,000 kilometres (1,250 miles), to the southern U.S.A. and Mexico where they spend the winter. In spring they migrate north to breed.

Animal migrations also occur in water. Grey whales spend the summer in the cool waters of the Arctic Pacific. They move down to the

Arctic Warbler

White Stork

Swallow

Swallow

Golden Plover (Pacific)

Arctic Tern

Wandering Albatross

warmer waters off California to breed. Fishes such as cod and herring also migrate long distances between their breeding and feeding grounds. The Atlantic salmon migrates between fresh water and the sea, but this is not a yearly journey. Salmon stay in fresh water until they are between two and eight years old. Then they make their way down to the sea where they spend several years. They return to fresh water to breed, always to the very rivers where they were born.

Migration is also a vital feature of the life cycle of amphibians such as frogs, newts and toads. They live most of their life on land, but each spring they return to the water to breed.

Some animals make occasional one-way mass movements to another area. These journeys are not migrations but emigrations. The Norwegian lemming is an emigrating animal. Every few years the lemming population outgrows its food supplies. When this happens huge hordes of lemmings move away. Most of them die but those that survive have enough food. Locusts also make periodic emigrations, again as a result of overpopulation and a consequent shortage of food.

Some of the main bird and butterfly migration routes of the world. The circles show where the travellers set out from, and the drawings of them are located at their winter homes. After a few months they make the return trip.

95

Fitting In

Earth is covered with the most wonderfully varied forms of animal life. Each one, from the tiniest single-celled organism to the mighty blue whale, is perfectly designed to fit its own habitat, or natural living place. The ways in which animals fit their habitats are known as adaptations. Some of these adaptations have taken millions of years to develop.

Deserts are some of the hottest places on

Above far left: The large ears of this Butana goat help it to lose heat. The animal lives in desert conditions for a large part of the year.

Above left: The underside of a Costa Rican flying frog showing its webbed feet with gripping pads.

Above: The long claws on a chameleon's feet are an adaptation which enables it to climb trees in search of its insect prey.

Left: A mole surfaces from its burrow. Note the spade-like front feet, suited to digging tunnels, and the close fur which keeps out the dirt. Because the mole does not need to see, its eyes are not highly developed.

Matching up

Each creature is specially adapted for its own habitat and no creature can change places with another, but often animals with similar ways of life have similar adaptations. This is known as convergent evolution, and it occurs among quite unrelated species.

For example, many of the Australian marsupial mammals look and behave in much the same way as placental mammals in other regions. The marsupial mole is adapted to live underground in the same way as the placental mole; the Australian numbat has the same digging claws and long snout as the anteater of other lands.

Earth. During the day the Sun beats down unmercifully, but at night temperatures drop sharply. Food and water are scarce and the shifting sand makes travel difficult. Yet the camel is perfectly adapted to survive desert conditions. Its hair protects it from extremes of heat and cold. Its hump is a fat store on which it lives when food is scarce, and it can store water in special cells in its stomach. In addition its long eyelashes protect it from sun and sandstorms, and thick-soled hooves prevent the camel from sinking into the sand.

Other animals too have adapted to desert life. There are many rodents, most of which can exist with very little water. One, the kangaroo rat, never has to drink at all. It gets all the moisture it needs from its food. Most desert animals have quite large ears: the fennec fox in particular has enormous ears. These can pick up the slightest sound, but they also help to disperse heat, in much the same way as radiators do.

By contrast, the Arctic fox has tiny ears. These help to prevent heat loss. Its thick fur coat also helps to protect the Arctic fox against extreme cold. The Arctic pika, a relative of the rabbit, is another creature built to survive the cold. It lives in the high

Himalayas and its short, stubby body and small ears help to prevent heat loss.

The mountainous regions of Europe and Asia are cold and inhospitable. The only forms of plant life on the high slopes are mosses, lichens and grasses. At these heights the air pressure falls and the atmosphere is rarefied. Many birds are at home in these conditions. They are well adapted to these altitudes, and their thick covering of feathers provides excellent insulation. The chamois, mountain goats and sheep, and the yak are also adapted for these rarefied conditions. They have thick coats of hair or wool, and being agile animals they can move easily up and down the mountain slopes.

There are many more examples of adaptation among the invertebrates. Butterflies that feed on nectar from deep-belled flowers have a long proboscis or sucking-tube to reach their food. Other insects have adapted to extreme conditions, and can be discovered in caves, in burning deserts, in fast-flowing streams, in the ground and high above it.

Animals have their preferred habitats. They also have their preferred food. Many herbivores (plant-eating animals) graze on the African grasslands, but because of their preferences there is very little competition. The giraffe, with its long legs and neck, feeds on leaves at the tops of trees. No other animal can reach that high. Various antelopes browse on lower shrubs and trees, and gazelles, wildebeeste (gnus) and zebras feed on the grasses. Even among these there is no competition. The zebra feeds on the coarse tops of the grass, the wildebeeste feeds on the leafy centre, and the gazelle browses on the shoots.

97

The Winter Sleep

Animals survive the rigours and discomforts of winter in various ways. Some animals, such as birds, migrate to warmer climates and a more plentiful food supply, but others actually sleep the winter away. This winter sleep is called hibernation, from a Latin word meaning to pass the winter. It takes various forms and is different from normal sleep.

In true hibernation, preparations start many weeks before winter begins. The animal eats a great deal and stores up reserves of fat on its body. As winter approaches, the animal finds a sheltered spot, perhaps a burrow or hole in the ground, and falls deeply asleep.

During its long winter sleep, all the animal's body processes are reduced to an absolute minimum. The animal's heartbeat slows right down, and its breathing becomes very slow and faint. Its body temperature drops quite sharply until it is only a degree or two above the temperature of its surroundings. In this state of almost suspended existence the animal uses up very little energy. It does not eat or drink but instead lives on its reserves of body fat. The animal sleeps in this way for week after week. Then, when spring arrives and the days are warmer, its temperature rises and it wakes up again.

Only a few small, warm-blooded mammals go into true hibernation and experience these dramatic body changes. They include bats, dormice, ground squirrels, hedgehogs, skunks and woodchucks. Even some of these animals wake occasionally on warm days, eat some food, and then go back to sleep.

Many other animals also sleep for part or most of the winter. Bears, badgers, rabbits and tree squirrels sleep for days at a time during cold weather, but it is not the deep sleep of hibernation. They emerge on warm days and either hunt for food or eat supplies they have stored.

Some birds and fishes become very drowsy during winter. Various fish such as carp and tench bury themselves in mud at the bottom of rivers and ponds during cold weather. If a tench is dug out of its winter bed, it is absolutely rigid and shows no sign of movement until tapped with a stick, when it wakes.

Among birds, the American whip-poor-will is the nearest to a true hibernator. It sleeps for most of the winter and its body temperature drops quite substantially from about 38°C to 20°C (100°F-68°F). The Indians knew about this habit of the whip-poor-will years ago, but scientists thought it was just a folk tale until 1946, when an ornithologist (a person who studies bird life) discovered some birds in hibernation. By ringing some of the birds, it

has been found that they go into the same rock crevices year after year for their winter sleep.

A number of cold-blooded animals, apart from fish, also hibernate. Snakes and lizards hide away in cracks in the rocks and holes in the ground. Frogs, salamanders and toads bury themselves in mud at the bottom of ponds to escape the frost, and many of them huddle together for extra warmth. Some species of snails also hibernate. They seal up their shells with a protective mucus, and some of them also hibernate in groups.

Most insects get through the winter in one of the early stages of their life cycle, a process that is not truly hibernation. Some pass the winter as an egg; others, such as various butterflies and moths, spend the winter in the larva stage as a caterpillar or grub hibernating among dead leaves. Some survive the winter as a pupa or chrysalis, safely enclosed in a cocoon. A few butterflies and moths spend the winter as adults. They take shelter in hollow trees or come inside houses, then hibernate.

Above: The dormouse, seen here still wide awake in a nest of leaves, gets its name from the Latin word *dormire*, to sleep, as in the boarding school dormitory.

Above right: Curled in a tight ball, a hedgehog hibernates among leaves.

Right: A mass of snails aestivating—having a summer sleep—on a fence post in Spain.

Siesta time

Just as some animals go to sleep to survive cold, so others go to sleep to survive extreme heat or drought. This process is called aestivation or summer sleep. It occurs among animals in the tropics, particularly in deserts and regions with a very dry season. Aestivation is exactly the same as hibernation, but in reverse —a sort of long-term siesta.

The African lungfish is one of the animals that aestivates. It lives in stagnant or sluggish rivers which evaporate in the dry season. When this happens the lungfish burrows into the mud of the river bottom. It breathes air through a narrow tunnel leading up from its burrow to the surface of the mud. The aestivating lungfish can survive up to four years in this condition. Other animals that aestivate include various frogs, toads and desert rodents.

Living Together

Animals do not form friendships in the same way as humans, but sometimes two unrelated species form a very close association. This is known as symbiosis, which means literally 'living together'. The association can be a type of partnership in which both species benefit, or it can be an association known as parasitism. In this relationship one creature lives entirely at the expense of another.

One of the most unlikely partnerships is between a small bird, the Egyptian plover, and the Nile crocodile. While the crocodile basks on the river bank, the bird feeds on leeches that infest the crocodile's body. The plover even 'cleans the crocodile's teeth', hopping inside the animal's huge mouth to remove leeches and pieces of food. Both animals benefit—the bird gets food and the crocodile is freed from the leeches.

The African tick-bird and the cattle egret often associate beneficially with larger animals in much the same way. They form partnerships with game animals such as antelopes and buffaloes, and feed on the ticks and small insects found on their bodies. Egrets feed on the grasshoppers disturbed by the movement of the animals. As an extra service the birds warn their partners of approaching danger.

A little bird, the honeyguide, forms a partnership with the ratel, an African member of the weasel family, often called the honey badger because of its habits and appearance. The ratel is fond of honey, breaking into the nests of wild bees to find its food. The honeyguide feeds on the bee grubs and wax, but cannot break the nests. A honeyguide which finds a nest then calls for a ratel. Once the ratel arrives it breaks the nests open for both of them.

Fish also form partnerships. Some fish, such as the tiny Red Sea wrasse, act as 'cleaners' for others. They pick off and eat fish lice and even patches of fungi and harmful bacteria from the skins of other fish. Many of the fish that they clean in this way are much larger and include sharks and the ferocious moray eel.

The remora, or shark sucker, is symbiotic with the shark. The remora has a large sucker on top of its head. It attaches itself to the underneath of the shark and travels along with it. When the shark feeds, the remora shares its food. In return it keeps the shark's skin free of parasites.

All these partnerships are examples of commensalism, which means 'eating at the same table'. Both species benefit from the association, but each can live independently. There is another, much closer, type of partnership in which each species is absolutely dependent on the other. This is known as mutualism.

Mutualism occurs among hermit crabs and sea anemones. Partnerships between these two animals are very common. The hind parts of the hermit crab are very soft and it usually lives with its abdomen inside a whelk shell. Almost invariably a sea anemone lives on top of the shell. The crab is protected by the sea anemone's stinging tentacles, and the anemone is carried around to fresh feeding grounds.

On a smaller scale, mutualism occurs among wood-eating termites and tiny protozoan animals called flagellates which live inside their intestines. The termites do not produce the necessary enzymes to digest wood. Instead the wood is broken down by the flagellates. In return the flagellates, which are unable to live anywhere else, are sheltered and fed.

In parasitic associations, one animal, the parasite, lives closely with another animal, known as the host. The parasite takes all its food directly from the host and gives nothing in return. The parasite usually weakens the host as a result, but rarely kills it.

Some parasites live outside their hosts. They are called ectoparasites. Usually they live by sucking the host's blood. Fleas, lice, mosquitoes and ticks are all ectoparasites. Many of them also carry disease. Fleas can carry plague, mosquitoes carry malaria, and lice can carry typhus.

Those parasites which live inside their hosts' bodies are called endoparasites. They include flukes, roundworms and tapeworms.

Right: Four sea anemones on top of a discarded whelk shell, in which a hermit crab is living. This kind of partnership is called mutualism.

Below: A crocodile surrounded by Egyptian plovers. These birds remove leeches from the reptile's body.

Below right: A fish louse clinging to its host. This parasite is not a true louse, but is a small crustacean. It holds on to its host with the aid of two suckers.

Keeping herds

Man is not the only creature to tend herds of other animals in order to get food. Some species of ants do this, too. The ants like to feed on honeydew, a sweet liquid produced by aphids such as greenflies. The ants 'milk' the aphids regularly and keep guard over them. Other insects are driven off by sprays of formic acid from the ants.

Some ants move their 'herds' from one plant to another to give them better feeding conditions, and one North American ant gathers aphid eggs safely into its nest to overwinter. In spring the ants carry the newly-hatched aphids out and set them on suitable plants.

Scavengers

Scavengers are creatures that feed on dead and decaying matter, both plant and animal. They play a vital rôle in nature. Some plants and animals are eaten alive or freshly killed; many others die from disease or old age. Even the larger predators, such as the lions, leave a certain amount of their kill behind. All this means a great deal of waste, which is where scavengers come in. In effect they are nature's waste disposers, clearing up dead and rotting matter which would otherwise accumulate to an impossible extent.

More importantly, scavengers also act as waste processors. Some of them break down rotting matter so that essential nutrients and minerals are returned to the soil. This keeps the soil fertile and helps in the growth of plants. The plants are then eaten by animals which in turn are eaten by other animals. Without scavengers this natural process of energy turnover and recycling of materials could not continue.

Some of the most valuable scavengers are insects, particularly beetles and flies. The well-named sexton or burying beetle is one. It helps to clear away the corpses of small animals such as birds and moles. Other scavenging insects include various water beetles; the dung beetle that eats and lays its young on animal dung which it buries below the soil; and the horn moth. This moth feeds and lays its young on animal horns, generally the very last remains of a kill. The wood-eating insects, such as death-watch and furniture beetles, are also scavengers.

Among the scavengers on a larger scale are hyaenas, jackals, vultures and crows. These animals feed mainly on carrion (dead animals), swooping down on the remains of a kill or waiting until a sick animal dies before beginning to feed. Some of them are also hunters in their own right.

There are three species of hyaenas—spotted, brown, and striped. The striped hyaena is found in India, the Middle East and North Africa. It is a true scavenger, getting most of its food from human refuse. The brown hyaena lives in southern Africa, while the spotted hyaena is found all over Africa and in south-western Asia. Hyaenas are known mainly for their scavenging habits, feeding on the left-overs from lions' kills. Their strong jaws and teeth can consume every part of a carcass, except the horns. The spotted hyaenas occasionally attack a wounded animal, or hunt in packs to kill larger animals. When this happens it is sometimes the lion that scavenges the hyaena's kill rather than the other way round.

Jackals are often found with hyaenas, scavenging the same kill. Jackals are swifter and more agile and are often able to grab more food. Like hyaenas, they also hunt their own prey, but they rarely attack anything larger than a young gazelle.

Vultures rarely kill. These huge birds are almost exclusively carrion eaters. There are two main types of vultures—the New World vultures of North and South America, and the Old World vultures of Africa and Asia. Vultures' bodies are specially adapted for carrion feeding. They have powerful, hooked bills which can easily rip skin and flesh. Their talons are quite feeble and are designed more for running and walking then for grasping prey. Most species have almost bald heads and necks which are suited to plunge deep into a rotting carcass. Vultures glide on large, broad wings. They are carried into the air on thermal updraughts (rising currents of hot air) and can remain soaring high in the sky for long periods. When searching for food, each vulture 'patrols' a specific area. When one vulture sees a carcass, it swoops down, its wing feathers producing a whistling sound. Other vultures notice its descent and a flock soon gathers. That flock can pick a donkey's carcass clean within 30 minutes.

All these animals are the largest of the scavengers, but at the other end of the scale, at the very heart of the natural recycling process, are millions upon millions of microscopic organisms. They include bacteria, fungi and protozoa (single-celled animals). They arrive in vast armies wherever there is dead or decaying organic matter. Generally they are known as decomposers. It is these decomposers that break organic matter down into its simplest form.

Right: The spotted hyaena is found all over Africa and south-western Asia. It stays in its lair by day and hunts at night, looking for carrion or easy prey.

Below: The black-backed jackal lives in Africa, from Ethiopia south to the Cape of Good Hope. It is an important scavenger.

Below right: Vultures gather round the body of an antelope.

Scavenging in towns

The growth of big cities and large, partly built-up urban areas has provided animals with new sources of material for scavenging—people's household waste. One of the results in northern Europe has been an increase in the number of foxes living in or close to towns, while in North America coyotes have also become 'city slickers'. These wild animals raid refuse bins in search of scraps of food. In some places badgers are also known to raid refuse bins.

Life at Night

When most people are glad to be in bed at night after a tiring day many animals who sleep during the day are at their busiest. There are several reasons why this should be so. Many animals come out at night because most of their natural enemies are not around, or because it is much easier to escape detection in the dark. Others emerge because their prey is out and about by night.

When the Sun has set the temperature of the air drops sharply. As it drops, the water vapour in it condenses to form dew—or frost in colder weather. In this cooler, damper period of the day animals whose bodies are likely to dry up, such as earthworms and many other forms of invertebrates, can move around. So out come the larger night hunters that prey on these creatures, such as shrews, and they in turn are hunted by owls and badgers.

Because of the lack of light, animals which hunt for their food at night tend to have differently-developed senses from those operating by day. Owls, for example, have larger eyes, and if you see a creature with very big eyes you can be fairly sure it is nocturnal —a creature of the night. The senses of smell, touch and hearing are also highly developed among nocturnal creatures.

One group of animals has developed an extra sense for night work: bats, the only flying mammals. Some bats, particularly the fruit-eaters, seek their food by day, but a very large number fly when there is little or no light, and they need some sense that will stop them from crashing into objects in their flight path. Yet a bat can fly neatly through a tangle of wires or branches. Bats can do this when there is definitely no light at all; scientists have tested their abilities in blacked-out rooms.

The bats find their way, and the night-flying insects such as moths on which they feed, by an echo-location system, which works something like the sonar used by Man to find the depth of the sea and detect submarines.

While flying a bat sends out a loud series of ultrasonic sounds—that is, sounds that are too high for human ears to detect. Young people can usually hear a few of the lower notes made by a bat, but they lose this ability as they grow up. The signals which the bat emits are bounced off objects in its path, as an echo. The bat's brain can detect one of these echoes within one-thousandth of a second of sending out the signal. With such a speedy response it can take avoiding action as quickly as if it could see.

The bat puts out its signals either through its mouth or its nostrils. A bat that sends out

Above: A painted reed frog calling at night in Natal, South Africa. Frogs make most of their mating calls at night.

Above right: Home with supper: a barn owl flies in carrying a shrew.

Far right: A fruit bat from Gambia in West Africa. The large eyes show that this bat is a nocturnal animal, though some fruit bats hunt for food by day.

signals through its nostrils usually has a leaf-shaped projection around its nose, which helps to focus the signals. A bat which transmits sounds through its mouth has a long snout. Night-flying bats feed very largely on moths, which are also creatures of the dusk and dark. Some moths can also send out signals, which 'jam' the bats' sonar, in much the same way as one radio transmitter can jam another. Yet bats can fly together in a huge swarm without jamming one another—and some of these swarms can be millions strong. A swarm of that size can be seen when cave-dwelling bats all take to the air at once with the approach of night.

A big swarm makes quite a loud noise, with millions of wings beating at once, but on their own bats fly silently. Most nocturnal creatures move very quietly indeed. Owls have downy fringes to their plumage, which help to muffle any noise of wing-beats.

Another night-flying bird is the nightjar, of which there are 67 species. They feed on insects which they catch on the wing, and even pick up a small songbird if it is out late. Like the owls they have large, lustrous eyes, but while their eyes are one each side of the head, like those of most birds, those of owls both face front like human eyes. This forward-facing vision gives the owls very keen eyesight, so they are able to detect and pounce on small ground-living creatures.

Time for a sleep

Most birds that rely on insects for their food emigrate to warmer climates with the onset of winter, but one species of nightjar, the whip-poor-will of western North America, spends the cold weather in hibernation. It is one of the few birds that hibernates.

The hibernating nightjars hide away in crevices in rocks.

Fishing bats

One species of hare-lipped bat lives almost entirely on fish. This bat has long hind legs and large claws. It lives near the sea in tropical America. Working at night, it skims over the surface of the water, combing the sea with its claws until it catches a fish.

Scientists do not yet know how the bat finds the fish, because its sonar system would probably not work in water. For one thing, the echo reflected off the surface of the water would be very like an echo from the fish.

Electric and Luminous Animals

Two of the most interesting groups of animals are those which produce an electric current, and those which can emit light. The electric animals are fishes, and there are two main kinds, those which use their current as a weapon in hunting, and those which use it for navigation and steering.

The best-known of the electric fish is the electric eel. This animal is eel-shaped, but not really related to other eels—it belongs to a group of freshwater fishes that includes carp and minnows. The rear four-fifths of the electric eel's body form a living battery, with layers of plates in it just like those of a car battery. The electric eel produces current at three different strengths. With the lowest voltage current it can find its way, as do many other fishes. The second current is stronger, and may be used to lure other fish towards the eel. Fishermen have lured tunafish and others into their nets with current of similar strength.

When the eel's victim is close enough, it receives the full-strength current—at an incredible 300-800 volts. A shock of this strength is enough to stun a horse, and would certainly knock a man unconscious. It renders the eel's victims helpless, and the eel eats them at once.

The electric eel lives in the Amazon and Orinoco rivers of South America, as do its relatives the knife-fishes. Several other electric fishes are found in the rivers of Africa, including another species of knife-fish, the elephant fishes and the electric catfish. The electric catfish can produce shocks of more than 400 volts.

The elephant fish is one of the electric fishes which uses its current only for taking its bearings. It produces a rapid, continuous current of 3-10 volts, forming around itself an electric field similar to the magnetic field produced around a bar magnet—a pattern you may well have seen demonstrated with the aid of iron filings. In some way the fish can detect any interference with this electric field, and can use this information in the same way that a ship's captain uses radar to find his way in the dark or fog. Like other electric fishes, the elephant fish is a sluggish creature with poor sight and hearing. Knife-fishes have a similar life-style to the elephant fish.

One electric fish found in the deep oceans is the electric ray. About 20 species, often called torpedoes, live in warm waters, and they can produce currents of up to 50 volts—enough to stun their prey.

Several species of animals produce light, including fish, insects, and some one-celled creatures. The most familiar are fireflies and glow-worms, also called lightning bugs in North America where many of them are found. Most fireflies live in warm places, but the European glow-worm is found in cooler places such as England.

A firefly produces its light by a chemical reaction. A nerve impulse triggers off a combination of six chemicals in the insect's abdo-

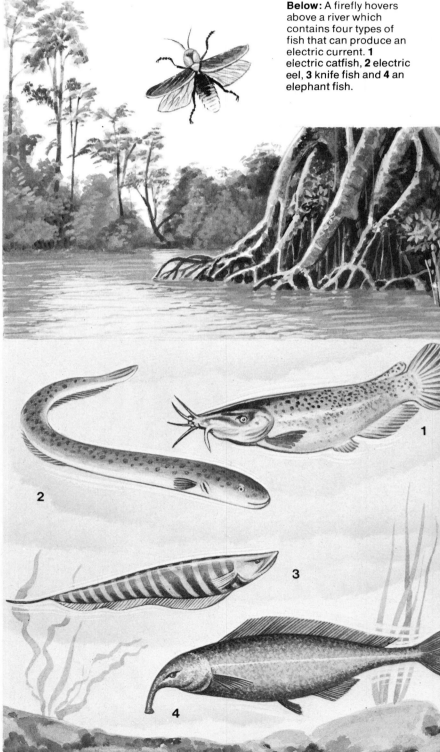

Right: The sea contains many luminous creatures. Shown here are **1** some plankton and shrimps, **2** the peculiar-looking angler fish, **3** hatchet fish, **4** lantern fish, **5** viper-fish and **6** a fish called *Photostomias guernei* which lives in the very deepest parts of the sea.

Below: A firefly hovers above a river which contains four types of fish that can produce an electric current. **1** electric catfish, **2** electric eel, **3** knife fish and **4** an elephant fish.

men, and a flash of light appears. The insect switches the light off by another nerve impulse which releases a seventh chemical, stopping the reaction. The insects use the light to find their mates, and each species has its own code of signals.

The light produced by fireflies and other creatures is heatless, and it is called bioluminescence, luminescence being the technical name for light produced without heat. By studying the ways in which animals produce this light, scientists have made lightsticks, which glow brightly for some hours when you snap them to mix the chemicals inside.

The sea is full of luminous creatures. The plankton of the surface waters of the warmer oceans often glow, particularly when they are stimulated by pressure waves. Such waves can be caused by underwater earthquakes, or by a boat or even a swimmer. Jellyfish and similar sea creatures produce luminescence, causing patches of light in the sea. Certain marine worms glow when disturbed.

More remarkable still are the many deep-sea creatures which produce light. They have been seen by scientists in underwater exploration vehicles such as the bathyscaphe. There are luminous shrimps, squid, and predatory fishes which shine their lights when hunting their prey. Some underwater creatures seem to use the lights as a form of communication, in a world which is otherwise pitch dark.

Fishing with light

The deep-sea angler fish goes looking for its food with a lantern. Part of its back fin is prolonged into a slender fishing rod, which extends forward in front of its nose.

On the end of the rod is a glowing light, which attracts other fishes. Some kinds of barbel fish have their lights hanging on the ends of what look like beards.

One species of viper-fish has its lights on the inside of its mouth. Small fish swim into this cavern of light, only to be swallowed.

Red and green

One species of firefly is called the railroad beetle because it produces lights which look somewhat like railway signals.

This insect, which is found in Paraguay, has a red light at each end of its body and green lights down the sides.

THE PLANT KINGDOM

If there were no plants in the world there would be no animals. Only plants are able to make complex living tissues from simple inorganic materials—water, carbon dioxide gas and mineral salts. Animals eat plants and other animals.

Plants are necessary for other reasons. They provide shade and shelter, and in deserts they may be the only stores of water to be found. Birds, bees, wasps, ants, termites, squirrels—all make their homes in trees, and larger animals may shelter among their roots. The teeming life of the soil depends upon the broken-down plant material that gives the soil its structure.

Even in the sea, plants are the source of life. Tiny floating plants, the phytoplankton, so small that they can be seen only with a microscope, trap the Sun's energy and provide food for microscopic animals, the zooplankton, which in turn are eaten by small fish, and so on up the food chain.

Plant life began in the sea, but today plants have conquered most of the land surface. Mosses and ferns remain restricted to wet places, but succulents such as cacti can live in the deserts. At the very edge of permanent snow and ice are the lichens, close partnerships between fungi and algae, which are themselves different forms of plants.

Ancient beech trees in the New Forest in autumn with bracken growing around them. The New Forest is one of the largest stretches of unspoiled woodland in southern England. It was created by William the Conqueror in the 1070s as a place to hunt, and has remained the property of the British Crown ever since.

Plant Groups

The plant kingdom includes a remarkable variety of living organisms, from giant redwood trees to seaweeds, slimy fungi and bacteria. Because plants do not easily form fossils, no one knows exactly how the different kinds arose and the relationships between them, so the family tree of the plant kingdom has to be based on rather sketchy evidence.

The most familiar plants are the flowering herbs and trees. Their bodies are made up of a stem which bears leaves above ground, and which continues into the soil as roots. They have a complicated series of conducting tubes running from the roots to the leaves, with separate tubes for water and food. The reproductive structures are flowers—male and female organs surrounded by whorls of special leaves; the colourful petals necessary to attract insects; and the sepals, which protect the flower while it is still a bud. Wind-pollinated flowers are not as big and colourful as others, but they are still enclosed in special leaves, the bracts.

Pollen grains, the male sex cells, are carried by insects or blown by the wind to the female structures, where they grow long tubes to the egg cells. The fertilised eggs of flowers are shed as seeds, each with two protective coats and a food supply. The ovary (seed box) wall is often adapted to help the seed dispersal. Other parts of the flower may also form the fruit structures. The flowering plants belong to the group called the Spermatophyta—the seed-bearing plants—and they form one of two classes in this group called the Angiospermae, which means seeds in a case.

The other class of the Spermatophyta is the Gymnospermae, which means naked seeds. Plants in this class include the conifers, cycads and ginkgoes. Instead of flowers they have cones—whorls of special leaves, sometimes woody, bearing either egg cells in egg-sacs, or pollen grains in pollen-sacs. The cones are either male or female. The plants rely entirely on the wind to carry their pollen from cone to cone. The seeds develop on the under-surfaces of the scales on the cone, without the protection of ovaries. The leaves of most gymnosperms are narrow and in some species even needle-shaped.

Ferns, horsetails and club mosses belong to the group Pteridophyta, meaning fern plant. They are much simpler in structure. Their water and food-conducting systems are not as advanced as those of the higher plants. When they reproduce they shed spores from the underside of their leaves or from cones. The spores are carried on the wind, and when they reach a suitable place they germinate, forming

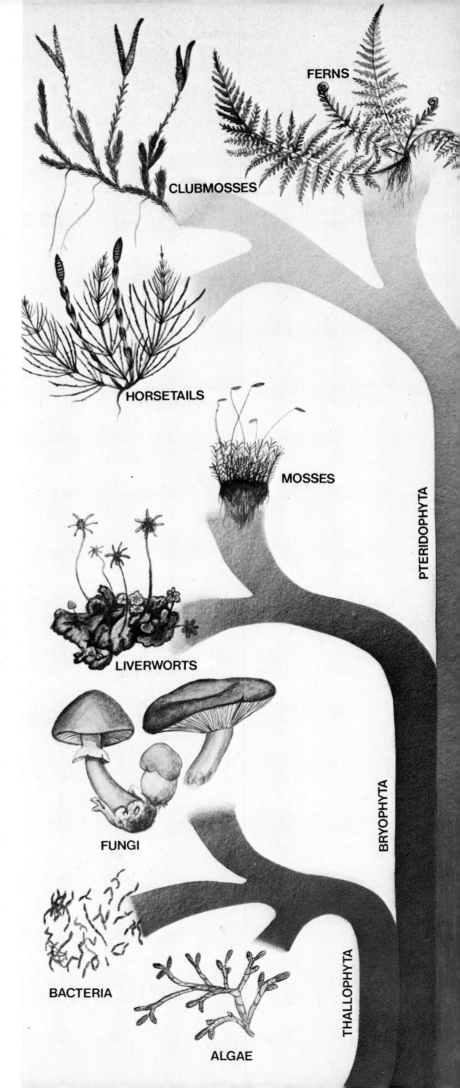

FERNS

CLUBMOSSES

HORSETAILS

MOSSES

LIVERWORTS

FUNGI

BACTERIA

ALGAE

PTERIDOPHYTA

BRYOPHYTA

THALLOPHYTA

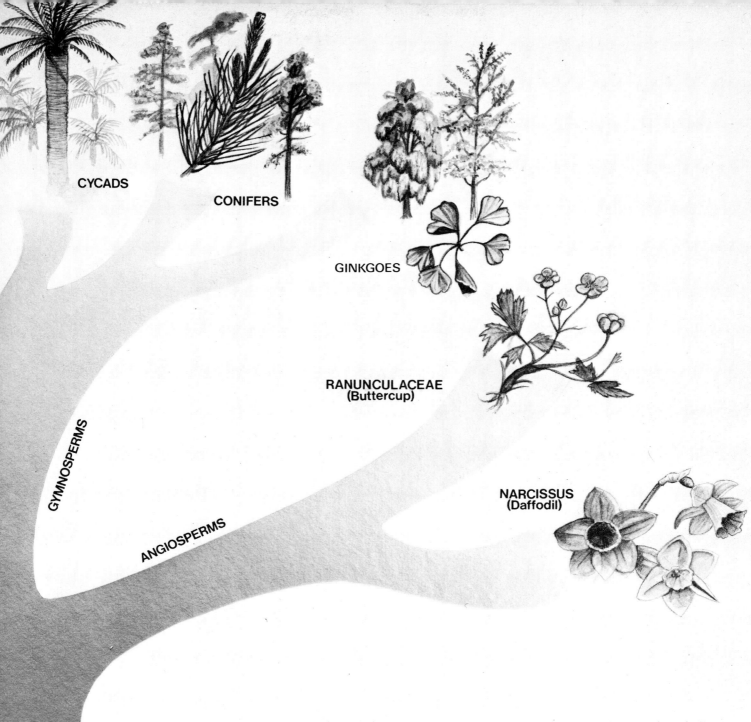

CYCADS

CONIFERS

GINKGOES

RANUNCULACEAE
(Buttercup)

NARCISSUS
(Daffodil)

GYMNOSPERMS

ANGIOSPERMS

SPERMATOPHYTA

a delicate plate of cells, the prothallus, on which the sexual organs develop. After fertilisation the new plant grows as a parasite on the prothallus, then becomes independent.

The mosses and liverworts belong to the Bryophyta, the moss plants. They are much more primitive, with no conducting tissues and no proper roots. Unlike the groups already described, their bodies are not covered with a waterproof layer preventing evaporation, so they can live only in wet places. The plant body may be a flat frond, as in the liverworts, or a leafy shoot, as in the leafy liverworts and the mosses. The sex organs develop on the plant's body, and swimming male cells fertilise the egg cells. Each fertilised egg then produces a spore case on a long stalk. The spores develop into new moss or liverwort plants.

Algae and seaweeds belong to the Thallophyta, which means that the plant body is a thallus, a simple structure without leaves, stem or roots. They range from single-celled plants to the giant kelps, and reproduce in a wide variety of ways.

The fungi are a group in their own right. They contain no green pigment, so are usually dull in colour. They feed on living, dead and decaying organisms. They disperse and spread by means of spores.

The bacteria are often included in the plant kingdom because they have cell walls and some of them carry out photosynthesis (see pages 112-113). Their cell structure is very primitive. Many of them feed like animals. Some biologists think they belong to neither the animal nor the plant kingdoms, but are in a third kingdom, the Protista.

Feeding and Growing

Plants have a very special way of feeding. Green plants make their own body materials from simple substances such as carbon dioxide taken in from the air, and water and mineral salts taken up from the soil. Their structure reflects their needs for these substances. A plant has roots, which are underground branching structures anchoring the plant and absorbing food from the soil. So that it can take in carbon dioxide from the air, it has a large area of leaves, spread apart and raised into the air by a stiff stem.

The energy needed to combine these simple materials and make the complex structures of a plant's cells comes from sunlight. The green pigment chlorophyll, found in leaves and sometimes in stems, absorbs light, and the leaves use the light's energy to make the cell compounds. This process is called photosynthesis, from Greek words meaning 'putting together with light'. The surface of the leaves is covered in the cuticle, a waxy layer, that stops them drying out. In this layer are small holes, called stomata, that let gases in and out. The stomata can be opened and closed, so that if there is no light for photosynthesis the plant can shut its doors, so to speak, thus not losing too much water by evaporation. The middle of the leaf blade contains loosely-packed cells, so carbon dioxide entering through the stomata can easily diffuse from cell to cell through the air spaces.

Large animals have a blood-circulating system of arteries and veins carrying materials round the body. Plants have two transport systems, one to carry water from the roots to the leaves, and the other to carry food round the plant. Running all the way from the roots, up the stem and into the leaves is a series of continuous fine tubes, called xylem vessels.

A simplified drawing of the intricate structure of a tree. The vascular bundle in the top picture (a branch) contains the xylem vessels, carrying water; the phloem vessels, carrying food; and the cambium layer which provides the cells to produce new wood. The third picture shows in a similar way how these vessels are arranged in the trunk. In the second picture is the wood as you might see it when the tree is cut down. The heartwood is 'dead', and no longer carries liquids. The bottom picture shows the inside of a root, greatly enlarged.

Below: The structure of a leaf. A large area of the leaf is used to trap sunlight. Most of the chlorophyll is in the upper surface, which tends to face the light. The inset circle shows something of the complicated structure inside a leaf.

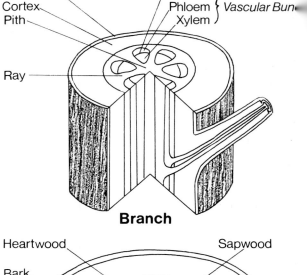

Branch

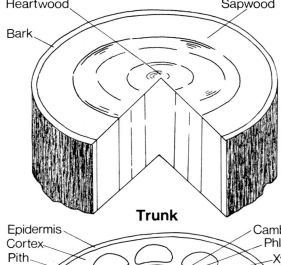

Trunk

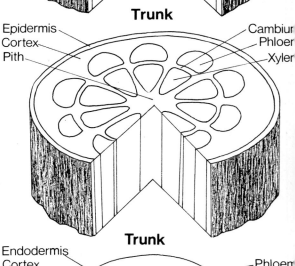

Trunk

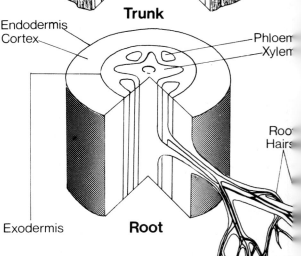

Root

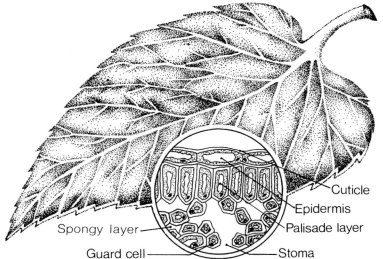

Water absorbed by the roots passes up these tubes to the leaves.

In the leaves some water is used for photosynthesis, and some is lost by evaporation into the air through the stomata, a process called transpiration. The evaporation of water from the leaves keeps the water moving up the xylem vessels. In the same way, when you suck a drink out of the top of a drinking straw more liquid is drawn up. The stream of water through the plant also contains dissolved minerals from the soil.

Because the roots are in the dark, they must rely on the leaves for their supply of food. Alongside the water-transporting system is another series of tubes, the phloem vessels. These carry food made in the leaves to the roots and to the growing parts of the shoots.

Many plants shed their leaves in winter and put out new ones in spring. During the summer they store food in their stems and roots, then move it up to the shoots in spring to help produce the new leaves. Food also needs to be transported to the growing flowers and fruit, the reproductive structures of the plant. The bundles of vessels transporting food and water are called veins. They extend to every

Above: A plant begins to grow. The fruit of a horse chestnut—a 'conker'—has split, and a long radicle is emerging.

Gas bubbles developing from the leaves of a water plant, a starwort, showing the activity going on inside.

part of the plant, branching into finer and finer strands in the leaves.

Unlike animals, plants do not grow equally all over. Only certain parts of the plant are able to grow. The tips of shoots, branches and roots are always growing. In woody plants there is new growth in the stem each year, producing woody tissue in the centre and a wider ring of bark on the outside. If this did not happen, the plant would be too tall and spindly, and the ring of bark would be too small to cover the expanding trunk.

Water for stiffness

Water helps to keep a plant rigid. Each plant cell has an elastic cell wall around it. The cell takes up water until the cell wall can stretch no more. The cell is then rigid—like a bicycle tyre when it has been filled with air, or a blown-up balloon.

On hot, dry days plants may lose too much water, and so then they droop. This shows how important water is for keeping them upright.

For large plants such as trees and shrubs this kind of support is not enough. They develop wood—rigid fibres which can support the great weight of their branches.

Plant Reproduction

Just as animals reproduce by laying eggs or producing minute copies of themselves which later develop into full-sized animals, so plants produce miniatures of themselves. Plants are not free to move around, so if conditions near the parent plant are not to get too over-crowded, there must be a way of transporting their offspring to more distant places where they can flourish.

The first step is to produce as small an offspring as possible. Many simple plants such as mosses and ferns produce single cells, each containing all the information needed to produce a whole new plant, wrapped up in a tough protective coat. These cells are called spores, and they are small enough to be carried away on air currents. When a spore lands in a suitable place it germinates—its cells multiply to produce a new plant.

The more advanced plants—the cone-bearers, which are mainly trees, and the flowering plants—produce seeds. Seeds are

Right: A typical flower, showing the arrangement of the flower's various reproductive organs.

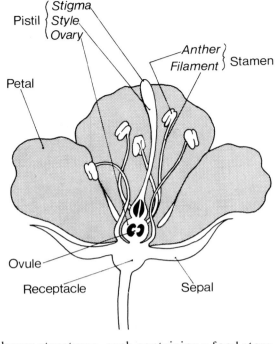

Below: The explosive fruit of the Mediterranean cranesbill—a wild form of geranium—scatters its seeds in all directions.

larger structures, each containing a food store to give the new plant a good start.

Just as animal babies are the result of mating between male and female animals, so plant spores and seeds are the result of male and female plant cells joining together. In this way, the characteristics of two different individuals are combined, producing a varied set of offspring, like human brothers and sisters, who seldom look exactly alike. The advantage of this is that some of these offspring may be better adapted than their parents to the conditions in which they live, and have an even better chance of survival.

This poses a problem, because male cells from one plant must be able to meet with female cells from another plant—and plants cannot move. Mosses and ferns, living in wet places, produce swimming male cells which travel through a film of water to the female cells. The cone-bearing plants and the flowering plants have developed pollen grains, male cells with protective coats. Pollen can blow on the wind and travel much greater distances to find female cells on other plants, a process called pollination. Pine trees even have pollen grains with air sacs to help them float on the wind. Because this is a rather haphazard method of pollination, conifers and wind-pollinated plants such as hazel and grass have to produce vast quantities of pollen to make sure some reaches female cells.

Many flowering plants have a more reliable method: insects or other animals carry their pollen. Flowers are very well adapted for pollination by means of animals. In the centre

of a flower is a round seed box containing the plant's egg cells. From the top of this box grows a stalk with a knob on top—the stigma. This sticky stigma catches the pollen grains. Surrounding the seed box are the stamens, a series of little anthers (bags) on filaments (stalks). The anthers contain the pollen.

Round the outside of the flower are the colourful petals that attract the notice of insects. At the base of the petals is nectar, a sugary solution rewarding the insects for their visit. As the insect feeds on the nectar, it brushes against the stamens, and the pollen grains get caught up on its hairy body. They are rubbed off on the stigma of the next flower the insect visits. Larger animals such as birds and bats also pollinate flowers.

Once the pollen grains land on the stigma they grow down to the egg cells and fertilise them. The fertilised eggs then develop into seeds—miniature plants surrounded by a tough coat.

The wall of the seed box, and sometimes other parts of the flower too, may help seed dispersal. It may form a wing, as in the sycamore, for wind transport, or a juicy fruit, such as a cherry, which can be eaten by birds. The seeds pass through the birds and out with the waste. The pods of peas and beans dry out until suddenly they burst explosively. The sepals, the protective leaves around the flower, form a hairy parachute for dandelion seeds. The fruit wall of burdock and goose-grass is hooked and catches in the coats of animals, which carry it away.

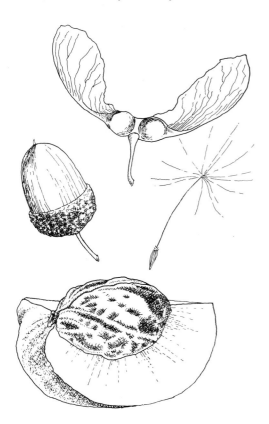

Above: The seeds of a dandelion float away on the wind, suspended from their silken parachutes.

Left: Some typical seeds showing the ways in which they are carried. At the top is a sycamore seed, with wings to help it ride on the wind; next comes an acorn, which falls straight to the ground, and is carried off by squirrels and other animals; and then the miniature parachute of a dandelion. The bottom picture shows a section through a peach, a juicy fruit which attracts birds.

Across the ocean

The hard shell of nuts is formed from the wall of the seed box. It protects the seed inside until it is ready to germinate. The largest nut, the coconut, has a fibrous shell, which traps a good deal of air inside it. For this reason coconuts float readily, and do not easily rot. So coconuts can, and do, float across the oceans from one place to another, where they are washed on to the shore and begin to grow. In this way the coconut has spread across the islands of the Pacific Ocean.

Flowering Plants

Flowering plants are found in almost every part of the world, from tiny cushion plants in the cool of the high Himalayas to giant cacti in the burning deserts. At first sight there appears to be an infinite variety of shapes, sizes and colours, yet there are some distinctive patterns.

Perhaps the most variable feature of any group of plants is its leaf form. Young leaves may be a different shape from old ones, and leaves on flowering stems are different from other leaves on the plant. Leaves developed in the shade are often larger and thinner than those developed in the sun. The arrowhead, a plant which grows on the margins of ponds and slow-flowing streams, shows the amazing variety of leaves that can appear on the same plant. Its underwater leaves are strap-shaped, allowing water to flow easily between them, while its leaves in the air are arrow-shaped.

The plant's genetics—that is, the built-in blueprint that governs how a living thing grows—determine the limits of leaf variation, but the details of each individual leaf's shape depend on the conditions in which the plant grows.

Leaves may be simple, with smooth or divided edges, like those of oak and maple, or they may be compound, made up of several small leaflets, like those of roses and palms. They may be shiny, like laurel, or hairy, like daisy leaves; evergreen, like pine, or deciduous (shed once a year) with beautiful autumn colours just before they fall.

Flowers on a plant do not vary like leaves. Each species or variety of plant has just one particular size, shape, colour and arrangement of its flowers. The flowering plants are divided into two major groups on the basis of their form and flowers. The larger group consists of the dicotyledons, so called because they have two seed leaves (cotyledons) in their embryos in the seed. They have leaves with the veins arranged in a net-like pattern. The parts of their flowers—petals, sepals, stamens —are arranged in groups of four or five, or multiples of those numbers. The dicotyledons include all the broad-leafed flowering plants. The simplest plant forms have the petals quite separate from each other, as in buttercups and magnolias. Flowers may have their petals joined together to form special structures such as tubes for insect pollination, as in campanulas and honeysuckle.

The smaller group, the monocotyledons, have only one seed leaf. Most have narrow leaves with parallel veins. Again, the petals may be free, or joined as in daffodils and arums. The monocotyledons include some of

The glowing colours of bell heather and yellow gorse attract insects.

The lines of light and shade running into the bells of autumn crocus help to lure an insect to the nectar and pollen.

Primroses have flat, open blooms, with clearly separated petals.

Opposite: The exotic bloom of a slipper orchid has a 'landing platform' on which insects can alight.

the most beautiful sculptured flowers in the world, the orchids, and some of the most inconspicuous, the grasses and rushes.

In both groups the flowers may be large and solitary, or small and grouped into showy clusters, called inflorescences. These clusters often look like single flowers, as in the daisies.

All the broad-leafed trees of the world are dicotyledons, since monocotyledons do not have woody growth. The largest monocotyledons are the palms, with fat stems thickened by long-lived leaf bases.

The range of forms of the flowering plants is extremely varied. They include some of the tallest trees in the world, the eucalypts; dwarf alpine and tundra plants; succulent cacti; tropical vines and lianas; thorny shrubs; clinging parasites; and feathery grasses.

Long and short lives

The lifespan of flowering plants may spread over weeks or centuries. There are four main groups:

Perennials live for many years, and most of them flower every year.

Biennials last for two years only. They grow in the first year, overwinter, and then flower and fruit in the second year.

Annuals grow, flower and fruit in one season, and overwinter as seeds.

Ephemerals have a very short lifespan and several generations are produced in one year.

Guide to flowers

Flowering plants are grouped into families containing plants that are basically similar in structure or habit. There are a great many of them: 45 in the monocotyledons, and more than 250 in the dicotyledons.

Some families contain a large number of species, others only a few. Among the most important large families are the Compositae (daisy, dandelion, lettuce, sunflower and marigold); Gramineae (grass, bamboo and the cereals); Cruciferae (cabbage, watercress, radish, turnips and wallflowers); Leguminosae (peas and beans); Rosaceae (roses, apples, plums, strawberries and pears); Umbelliferae (carrot, celery, hogweed, parsnip, parsley, dill, fennel).

Flowers of Mild Climates

The temperate regions of the world—those with warm summers and cool or cold winters—have many different types of plant life, depending upon the height of the land, the kind of soil and the distance from the sea. Much of these regions has been affected by Man, and around his settlements are fields of food crops and pasture for animals. The grassy meadows grow varied wild flowers, most of them able to survive being grazed by cattle and sheep, but some taking refuge in the hedgerows. Plants such as buttercups, dandelions, parsleys and poppies produce many seeds before they are eaten. Some, such as yellow primroses and cowslips and the starry stitchwort, flower early in the season. Others, including the purple orchids, the nodding snake's head fritillary and the globe flower, store food in their roots or other underground parts. Meadows and hill grasslands in summer are the homes of literally hundreds of different wild flowers.

Where the ground is more marshy, and along the edge of ponds and streams, the marsh marigolds or king cups make a blaze of yellow in spring. Later in summer the meadowsweet scents the air with its fluffy white blossoms. Many plants grow in the

Right: Corn marigolds, which are wild flowers, in a field of oats, a crop planted by Man.

Below: A typical forest of oaks, with a carpet of ferns and grasses, found in many parts of the temperate zone in the northern hemisphere.

water, putting out flowering heads into the sunshine. Carpets of white water crowfoot cover slow streams, and at the edges the arrow-shaped leaves of the arrowhead point towards the sky. The floating three-petalled frogbit flowers emerge, and further north the slender spikes of water lobelia grow like grasses from the water. White and yellow water lilies float their round leaves across the water, duckweeds crowd the water between, and pondweeds fill the water just below the surface. Bulrushes, reeds and sedges stand like soldiers along the shore, and willows and alders shade the banks.

Between the meadows are broad-leafed woodlands, mostly planted by Man. Only in the heart of the continents are the remnants of the original forests to be found. These contain many different trees—oak, ash, beech, lime, sycamore and birch, with hornbeam and chestnut in Europe, and hickory, maple, sweet gum and tulip trees in North America. In South America the main trees are beech and its relative, lengue, and in Australia eucalyptus forests have developed in the dry climate. The structure of these forests depends upon which species are most common in a particular place.

In oak and ash forests the leaves let through plenty of light, and there is at least one layer of shrubby undergrowth with hazel, holly, brambles, wild roses and honeysuckle, and a rich ground layer of herbs and ferns on the woodland floor. Many woodland flowers use food stored in underground organs to enable them to grow and flower early in spring before the trees come into leaf to shade them. Such plants include the attractive snowdrops, bluebells, celandines and wild daffodils of northern Europe, and the trilliums or wood lilies of North America. Rotting tree trunks are dotted with fungi, and many live trees have a coat of ivy, hanging on by sucker-like roots as it climbs towards the light. Plants of the ground layer include parasites living on the roots of trees, such as the toothwort, a parasite on elm and hazel roots.

Beech trees cast a deep shade and produce a thick layer of acid litter, so beech woods have little undergrowth. Wood anemones, wood sorrel, bluebell and hollow corydalis carpet the ground early in spring, but there are few flowers later in the season, except the May lilies of European forests.

The pine woods of northern regions have a flora of lichens, grasses and berries—such as crowberry and bilberry—and occasionally purple or white heather.

The coasts have a plant life all their own. Sand dunes are often held together by the whip-like marram grasses. Plants growing there must be able to survive being buried by

Above: Butterbur, a creeping spring flower of northern Europe which is frequently found near water.

Left: The flower of the teasel, a thistle-like plant which is native to Europe but is grown in North America. The dry flower heads are used to raise the nap on cloth.

blown sand. Colourful dune pansies and lady's bedstraw grow there, and sand sedges march like armies across the sand, their straight underground stems sending up shoots at regular intervals. Many coastal plants have thick, waxy leaves which reduce water loss. They include shiny sea sandwort, grey-coated sea holly and sea kale.

The mud flats are colonised by the succulent little glasswort, or by rapidly-spreading cord-grass, and the marshes behind are the home of such plants as pink sea thrift, purple 'sea lavender and sea asters.

Right: The wild rose, the ancestor of most of our garden roses.

Flowers of the Tropics

The climate at or near sea-level in the tropics is the best climate on Earth for plant growth. The year is one long growing season, with no dark, cool winters or wet and dry seasons. There is plenty of sunlight all through the year, rain every day and a warm temperature. As a result tropical forests have lush growth, layer upon layer of plants trying to dominate each other in their struggle towards the Sun.

The top of the forest is a leafy canopy formed from the crowns of close set trees. So tall do they become in this everlasting growing season that the soil is not deep enough to accommodate their roots, and many of them have developed aerial roots trailing from high branches to the ground to act as props. The most notable is the banyan, which can support a crown up to 660 metres (2,150 feet) in circumference, propped up by as many as 320 root-trunks for extra support.

There is never any shortage of water in the tropics, so most of these forest trees are evergreen, photosynthetising all year round. Some reach heights of over 80 metres (260 feet) in 300-400 years. The rainfall can cause problems, however. If these trees are to draw minerals up from the soil they must transpire (give off water-vapour), and wet leaves do not transpire quickly. Some figs and other species have developed 'drip-tips'—long pointed tips off which the rain runs easily.

Because tree trunks are in no danger of drying out in such a humid place, the bark of tropical trees is quite thin. Some trees, such as

Below left: The young pods of the cacao plant, from which cocoa is made. Cacao grows in tropical America and West Africa.

Below: The orange trumpet vine is also known as the flame thrower or firecracker vine. It is found in tropical America.

the cacao (cocoa) tree, produce flowers and fruits straight out of their trunks.

In such a crowded habitat there is great competition for light. Many plants rely on the support of others. By scrambling, clinging or twining they save on body materials and still reach high into the forest. These climbers include the rose-mauve bougainvillaeas, with hooked thorns; the abundant twining Dutchman's pipes; the blue trumpets of the morning glories; and many figs and vines. Lianas hang like ropes from the canopy, often twining around each other for extra support.

With so much lush greenery above it the forest floor is in almost perpetual shade. Few plants struggle through its deep layer of rotting leaf litter. Among them are the wild ginger and its relatives. Some plants do not need light because they feed as parasites on the roots of others. All they show above the leaf litter are their flowers—brilliant oases of colour, with no foliage. The most dramatic of these is the scarlet rafflesia flower of Sumatra, up to 1 metre (3 feet 3 inches) in diameter. The forest floor is a paradise for fungi of all shapes and sizes, which help to break down the leaf litter.

Tropical swamps are common in regions with so much rain. The water surface is hidden by layers of leaves—the stalked leaves of lotus flowers and the round flat leaves of water lilies—some, such as *Victoria regia* of the Amazon, reaching a diameter of 2 metres (6 feet 6 inches). Many tropical swamps and rivers are full of blue water hyacinths, whose large leaves completely fill the water. Some familiar house plants, such as philodendrons, come from tropical swamps.

Below: The parrot's beak flower, so-called because of its shape, is one of the many bright colours of the tropics.

Below right: The lotus is one of the most lovely of tropical blooms. The people of India, Egypt and China regarded it as sacred. This specimen comes from the East Indies.

Roots in the air

Many tropical plants do not put roots down into the soil. Instead they grow on the trunks and branches of trees. They put out slender roots to absorb moisture from the air,. and get their minerals from the host branches.

Plants of this kind are called epiphytes, a term which comes from Greek words meaning on a plant. Epiphytes include species of orchids, selaginellas (fern-like club-mosses) and Spanish moss—not a moss but a member of the Bromeliaceae, related to the pineapple. Certain bromeliads have cup-like rosettes of leaves which form miniature ponds. In these ponds you may find a variety of animal life—even tiny frogs.

Flowers of the Desert and Tundra

Lack of water is probably the most difficult problem that plants have to overcome. Not only do they need water to build their tissues and keep them stiff, but they also have to keep replacing the water they lose through their shoots by evaporation. No plants are totally waterproof. To replace this water they must have a regular water supply.

In deserts plants do not get regular water. Not only is there very little rain, but what there is falls at irregular intervals, and in very varied amounts. Surprisingly, the deserts with the lowest rainfall are not the sandy wastes of the Sahara or Central Australia, but the Poles, where most of the moisture remains frozen and unavailable. The main cold desert area is the tundra, a region in northern America and Asia where the soil—except for the top few centimetres—remains frozen all the year round. This frozen soil is called the permafrost. High mountain tops are also deserts, not because of low rainfall, but because the strong, frequent winds dry everything up rapidly.

The word 'desert' usually conjures up pictures of sand or rock dotted with exotic cacti and thorny shrubs, and many deserts are like this. Cacti grow in the deserts of North and South America. They and their African equivalents, the euphorbias, get round the water problem by storing it in their succulent (juicy) tissues. For this reason they are often called succulents. Cacti provide some of the most eye-catching flowers of the desert—large, brilliantly coloured, with names such as strawberry cup and claret cup. Other names, such as barrel cactus, describe the shapes of these water-storing plants.

Cacti do not have leaves, through which they would lose water fast. Instead they photosynthetise with their green stems. These stems are well waterproofed with a shiny wax coating, and are armed with spines against thirsty desert animals. Underground the cacti have wide, shallow roots, which extend for many metres around them to trap as much rain from the occasional storms as possible before it evaporates. A giant saguaro cactus 15 metres (50 feet) tall may have roots just as long. Other plants, such as the night-flowering *Cereus*, have their reservoirs underground.

Some desert plants—for instance the bean caper—are more adaptable. They have normal leaves when water is plentiful, shed them when it gets too dry, then grow more leaves after rain. The sagebrush's leaves get smaller and smaller as the drought progresses. Such trees as the mesquite and tamarisk send deep roots down to the permanent water far below

the desert, and stand out as lush green dots in a burned-out landscape.

Some of the most attractive desert flowers live underground during the dry season. They store the previous year's food in bulbs, swollen stems or roots, ready for rapid growth as soon as some rain comes. These plants include the tulips, irises and lilies so often grown in gardens elsewhere for their beauty. Other plants, such as daisies, poppies and verbenas, hide as seeds. After rain they sprout rapidly, and grow, flower and produce fruit in just a few weeks. They cover the desert in a carpet of blossoms after a shower.

A striking feature of desert plants is how far apart they are. Creosote bushes, for example, grow a long way from each other in the middle of a desert, but are closer towards the edges where rain is more frequent. This shows how fiercely the roots compete for water.

Tundra and alpine plants cannot store water because it would freeze, but they have many of the same features as hot-desert plants. Many heathers have small leaves with thick waxy coverings. Woolly plants are common, such as the edelweiss of the Alps and the espeletias growing in the high Andes. The woolly covering traps a layer of air next to the plant and so helps to reduce evaporation.

Above left: The woolly covering of the espeletias of the high Andes helps to reduce evaporation.

seed in the tundra conditions, so they have adopted other ways of reproducing—side bulbs, and buds that drop off and grow into tiny new plants. Lichens are among the best survivors. They can be found in places where there is too little soil, too little water, and too much or too little heat to support any other form of life.

Many plants, such as the moss campion and dwarf geraniums, form low cushions. By staying close to the ground they escape the worst of the wind. Others, including the stonecrops and saxifrages, form tight rosettes of leaves close to the ground. As in the hot deserts, there are many plants with bulbs and swollen stems, such as irises, hyacinths and crocuses. Some flowers literally follow the Sun, pointing their centres at it and following it in a circle through the day.

Tundra plants have an extra problem: the short growing season. Many have adapted by becoming slow-growing dwarfs, so they do not make heavy demands on their environment. Others cannot manage to flower and set

Above: Tall cacti growing in a semi-desert environment in the Galápagos Islands.

Plant that comes to life

One of the most interesting desert plants is often known as the resurrection plant, because it comes to life again after being apparently dead. Its other name is the rose of Jericho, and it grows wild in Arabia, Egypt and Iran.

When the seeds are ripe in the dry season, the plant loses its leaves and curls up into a loose ball (top picture). In this form it blows hither and thither across the desert until the rains come. Then it unfolds (picture above) and becomes green again, and in a few weeks can grow to 200 millimetres (8 inches) high and produce flowers and hundreds of fruits.

Flowering Trees

The broad-leaved trees, members of the Angiospermae, are found in all but the most hostile corners of the world, and are just as varied as their relatives, the smaller plants we call herbs. Herbs and trees have the same kinds of flowers, and many are members of the same families. The only real difference is that trees can produce woody tissues and grow tall, and herbs cannot. Most broad-leaved trees are deciduous—that is, they shed their leaves every autumn, and grow new ones again in the spring. A few, such as rhododendrons, are evergreens like the conifers.

To see the diversity of the flowering trees, a close look at some tree members of the pea family, Leguminosae, is worth while. Laburnums, with their chains of yellow pea-like flowers, are easily recognised as relatives of the pea in late summer, when they have dangling pods of round seeds. Wisteria is a climbing member of the family which has the distinction of being the largest flowering plant in the world, with branches up to 150 metres (500 feet) long and 1,500,000 blossoms in a season. Broom is a shrubby member that has almost lost its leaves, and gorse has green spines instead of leaves.

The flamboyant-tree (also called the flame tree) of Madagascar is spectacular, with brilliant scarlet flowers that appear before its fern-like leaves. The Barbados pride also has flowers quite unlike those of peas, with long slender stamens protruding far beyond the spreading petals. The stamens of the *Calliandra* flowers are so long that the flowers look like fine powder-puffs. This gives the plant its popular name, powder-puff tree. The pea family also includes the evergreen acacias or wattles, with their thorny branches and fluffy catkins. All this variety occurs in a single family.

Flowers which appear before the leaves are very conspicuous, such as the blossoms of some flowering cherries. Large inflorescences (bunches of blooms) can also be impressive, like those of the jacaranda, which from a distance appears to have deep blue foliage when in flower. The largest inflorescence is that of the puya, up to 2.4 metres (8 feet) across. Some single flowers can also be large; the magnolias produce huge flowers with many whorls of creamy white petals.

Flowers are not the only colourful parts of trees. The spindle tree has bright pink and orange fleshy fruits, which remain on the tree long after the leaves have fallen. Kapok has silk-tufted fruits, which are used for stuffing cushions and mattresses. Some dogwoods have red shoots, buds and leaves. Many

deciduous trees such as maples have glorious autumn colours of red, gold, yellow and orange, but some have colourful foliage during the rest of the growing season. Copper beeches have deep red-brown foliage and weeping willows are delicate yellow-green.

Some birches are renowned for the colour of their bark. Silver birch has gleaming white bark, and the yellow birch a bronze-yellow trunk. Plane trees have peeling bark that reveals patches of tender yellow and green layers below. Other trees have aromatic oils that give them a characteristic smell. The blue gums (eucalyptus) are so called because they give off so much vapour that it produces a blue haze in the sunlight.

Trees exist in a surprising range of shapes. One of the most bizarre is the baobab tree of Africa, said to have been planted upside down by the Devil. It has a huge, swollen bottle-shaped trunk and stumpy branches that look more like roots than shoots.

Some examples of trees
and their blossoms (inset)
Above far left: Chestnut.
Below far left: Apple.
Left: Laburnum.
Below left:
Rhododendron.

Angle on shape

The angle made by the branches with the trunk has a dramatic effect on the overall shape of a tree. A very steep angle produces a thin tree, such as a poplar. A very wide angle can produce a huge spreading crown of branches, such as that of the banyan tree.

Drooping branches produce so-called weeping trees, such as the weeping willow, whose branches may reach the ground.

Tree products

Numerous flowering trees bear useful fruits—citrus fruits, cacao, coconuts, dates, carobs, olives, grapes and figs. Some trees produce a thick juice which is tapped off to make rubber.

Bark can be used to tan leather, and that of the paper birch was used by the American Indians to make their canoes.

Grasses and Vegetables

Grasses cover large areas of the Earth's surface. The pampas of South America, the savannahs of Africa, the steppes of Europe and Asia and the prairies of North America are the world's greatest grasslands. Ever since they first appeared on Earth grasses have been important food for animals—the bison that roamed the prairies and steppes, the llamas and guanacos of the pampas, and the vast herds of buffalo, zebra, wildebeeste, springbok and gazelles of the African plains.

The reason for grasses' survival under such heavy grazing is their ability to grow new shoots, 'tillers', from the bases of grazed shoots. Even after fire the underground parts of the grasses can soon put out new green growth, and the great numbers of tiny wind-blown seeds can quickly recolonise bare soil.

Man, too, makes great use of the grass family. His domestic animals—cattle and sheep—graze continually to produce milk and meat. Many staple food crops are the grains (seeds) of grasses such as wheat, rye, barley and oats. Wheat grain ground into flour is the main ingredient of bread. Rye is grown in the cooler parts of the world and on poor soils. Its flour produces a dark bread with a strong taste. Young plants are used as fodder for animals. Barley is used in beer, soup and stews, and the plants are used to feed live-stock. Oats, too, provide fodder, and their grains are used in porridge. The husks are an important part of the chemical furfural, used in refining oil and making nylon, dyes and antiseptics. In dry areas and on poor or waterlogged soils, millets are grown for hay and silage (stored animal fodder), and in tropical areas they are grown to make a kind of porridge. Millets can be grown right up to the edge of the Sahara.

Rice is a member of the grass family, and so is maize, which is a staple food for millions of people in Africa and America. Other grasses include sugar-cane and bamboo, one of the tallest grasses known and capable of growing 410 millimetres (16 inches) a day. Marram grasses are important for covering sand dunes to prevent them shifting. Dried grasses are often used as home decorations, and have been represented in artwork from ancient times.

Most of our vegetables are parts of flowering plants, often developed into a great many varieties by cultivation. This is especially striking in the genus *Brassica*, various species of which are valuable vegetables: *Brassica oleracea* varieties include cabbage, kale, Brussels sprouts, cauliflower, broccoli, calabrese and kohlrabi; *Brassica chinensis* is the

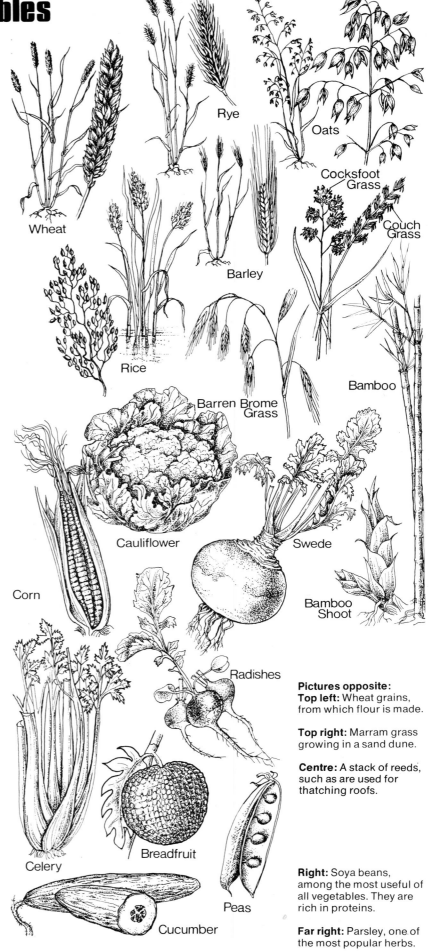

Corn
Wheat
Rye
Oats
Cocksfoot Grass
Couch Grass
Barley
Rice
Bamboo
Barren Brome Grass
Cauliflower
Swede
Bamboo Shoot
Radishes
Celery
Breadfruit
Peas
Cucumber

Pictures opposite:
Top left: Wheat grains, from which flour is made.

Top right: Marram grass growing in a sand dune.

Centre: A stack of reeds, such as are used for thatching roofs.

Right: Soya beans, among the most useful of all vegetables. They are rich in proteins.

Far right: Parsley, one of the most popular herbs.

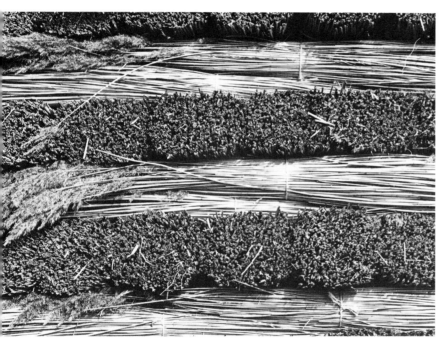

Chinese cabbage or pak-choi; *Brassica pekinensis* includes pe-tsaw, wong bok, chihli and spring greens; *Brassica rapa* is the turnip; *Brassica napus* is swede and rape; and *Brassica nigra* is black mustard.

Different parts of many plants are eaten as vegetables. Leaves are used from cabbage, chicory, endive, lettuce, spinach and chard. Underground leaf bases are eaten from leeks; and onions, shallots and garlic are bulbs —swollen leaf bases. Whole shoots of water chestnut, bamboo seedlings and asparagus are eaten, and the flowering shoots of broccoli and globe artichokes. The swollen tubers form some vegetables, such as potatoes and Jerusalem artichokes. Cassava (the source of tapioca), is grown for its rootstock, and celery for its fleshy stems.

Root vegetables contain starch. Carrots, radishes, parsnips, turnips, swedes, beetroot, salsify and celeriac are all swollen storage roots. The roots of sugar beet provide an important source of sugar, the pulp after extraction is used to feed livestock, and the tops can also be used as fodder.

Many so-called vegetables are really fruits. Breadfruit and jackfruit are cooked as staple foods in some parts of the world. Avocado pears contain one large stone. Cucumbers, marrows, squashes and gherkins are juicy fruits of the same family. Green and red peppers, tomatoes and aubergines all have little pips that are really seeds.

The most nourishing group of fruit vegetables is that of the legumes, members of the pea family. They provide valuable sources of protein. Some are used as dried seeds which are easy to store—haricot beans (source of Boston baked beans), lentils and the pulses, which include chick peas, pigeon peas and cow peas. Peas, groundnuts, butterbeans and broad beans are all eaten as seeds.

Flowers without petals

Grasses form the family Gramineae, and they are monocotyledons (see pages 116-117). Their flowers often go unnoticed because they are dull-coloured without petals or scent. It is these flowers that form the seeds of the grass that we harvest as grain. These seeds are dispersed by the wind.

The grains of wild grasses are mostly small, but those Man uses for food have been cultivated to produce large grains and very much heavier crops. New varieties are constantly being produced.

Conifers

The conifers or cone-bearing plants are some of the most successful plants in the world today, and are important parts of the vegetation in the cooler parts of the world. They hold some of the most spectacular plant records. The tallest tree in the world is a coast redwood in California 111 metres (366 feet) high, and the second tallest species of tree is a noble fir. The tree occurring at the highest altitude above sea-level, 5,000 metres (16,400 feet), is a silver fir, and the most isolated tree in the world was a juniper in the middle of the Sahara. Ginkgoes are the oldest kind of trees in the world, having first appeared on Earth 180 million years ago. The oldest living tree in the world is a bristlecone pine in Nevada, estimated to be about 4,900 years old.

The characteristic feature of the conifers is the reproductive structure, the cone. Each cone consists of a whorl of spore-bearing leaves, sometimes rather like woody scales. Cones are usually either male or female (rarely both), and male and female cones may occur on separate trees in some species. On the back of each scale of a male cone is a small bag of pollen. On the female cone's scale-leaves, often protected by a second scale leaf, are small round bumps containing the egg cells. The wind carries the pollen of conifers to the egg cells. After fertilisation the scales of the cones often close up and protect the developing seeds. When the seeds are ripe, sometimes after two or three years, the scales open wide and the seeds fall out into the air. Some have a papery wing to help them float.

Not all the fruiting structures look like the familiar pine cone. The yew produces a fleshy red cup around the seeds which attracts birds to carry them away, and the juniper produces blue-black berries.

The most typical conifers are the ever-

Far left: The sequoia or coast redwood called the 'General Sherman' is the world's largest tree in terms of the amount of timber it contains. It is about 3,500 years old.

Left: Flowers of the European larch. The pink are female, the yellow are the pollen-producing males.

Right: A detailed drawing of the cone of a Scots pine.

greens, such as pine, spruce and fir. They have dark green needle-like leaves with thick waxy coats, arranged spirally up the stem or in short tufted shoots. The yew and its relatives have flattened leaves, while the spiky monkey puzzle tree has thick overlapping triangular leaves, rather like armour plating. Some Australian conifers have succulent leaves. Not all species are evergreen—the larches shed their leaves in winter.

Some conifers form a conical-shaped tree. Others lose their lower branches and have tall reddish stems with a crown of branches at the top. Many have drooping branches, but the monkey puzzle tree has well-separated stiff branches forking in a distinctive pattern.

Cycads

The palm-like cycads have been in existence for millions of years, and are often called living fossils. They have thick stems covered in leaf scars and a crown of fern-like leaves, rather like feathery palm trees. Some species are as tall as 10 metres (33 feet). They can be very slow-growing, and some specimens may be several hundred years old.

Cycads occur mainly in the tropics, especially in Australia and Malaysia. They are rather more primitive than the true conifers. Although they produce pollen grains, when the pollen arrives on the female cone it releases male cells with a spiral of flagella (whip-like hairs), which swim towards the female cell. Some of these male cells are the largest sperms known.

Ginkgoes

Ginkgoes or maidenhair trees are survivors of an ancient group. They have long, soft male cones, rather like catkins, and female cones like long-stalked, very small green acorns. After pollination the female cones develop into dangling yellow plum-like fruits, each containing a single seed. The seeds eventually drop to the ground and germinate.

The trees are famous for their delicate foliage. The rough knobbly shoots produce fan-shaped pale green leaves with parallel veins. Each leaf has a notch in the middle, like those of the maidenhair fern.

Right: A Scots pine, which grows in northern Europe and Asia.

Left: The catkin-like male cones of a maidenhair tree, with new spring leaves.

Far left: The palm-like appearance of a cycad. Cycads grow only in a few parts of the tropics.

Left: The developing seeds of a cycad tree.

Right: A Cedar of Lebanon. These trees grow in groves on mountains in Lebanon and Turkey, and it was from their wood that Solomon built his temple.

129

The Fern Group

Ferns, horsetails and club-mosses belong to the group of plants called the Pteridophyta, which includes some of the oldest-known land plants. At the time when the coal we burn today was being formed, some 350 million years ago, the land and swamp vegetation was dominated by giant ferns and horsetails and their relatives. Today there are still some giant tree ferns up to 20 metres (65 feet) high in the tropics, but most of the common pteridophytes are much smaller.

The word 'fern' conjures up pictures of fronds, branching tufts of feathery leaves on woodland floors, cliffs and old trees. The fronds often spring from creeping underground stems, but many ferns do have proper root systems. Like the flowering plants, the fern plants have an outer waterproof skin, and a complicated system of internal conducting tubes for water and food.

The reproduction of ferns is very different from that of flowers. All the members of the Pteridophyta have spores for reproduction and dispersal. Each spore is just a single cell, with a very limited food store, and a cell wall that is flimsy compared with the coat of a seed. The chance of a spore surviving to grow into a new plant is very slight, but ferns produce vast quantities of spores, some of which survive. Spores are light enough to blow away on the wind, and in this way ferns can invade new areas.

There are two distinct stages in a fern's life-cycle. The one you usually see is the second stage, the fern plant itself. Fern fronds are mostly very branched, and they vary greatly in appearance. Young fronds are usually coiled in a tight spiral to protect the delicate growing point as it pushes up through the soil. Mature fronds may be feathery, or they may be quite undivided, like that of the hart's-tongue fern. The fronds of some ferns, such as the polypodies, have wavy edges; others, such as the maidenhair fern, have delicate rounded lobes. *Marsilea*, an aquatic fern, has four leaflets on top of a long stalk and looks like a four-leafed clover. *Azolla*, a tiny floating fern, has overlapping leaves, while the adder's-tongue fern produces a single spoon-shaped leaf each year, enclosing a tall spike of spore cases.

The spores form the earlier stage of the fern's life. They are produced in stalked spore cases on the undersides of the fronds. These cases grow in clusters, called sori, usually protected by flaps of skin until they are ripe. When they are ripe, the spore cases open explosively in dry weather, shooting out the spores into the air currents. Each spore grows

130

into a tiny heart-shaped plate of cells, the prothallus. On this prothallus the fern's sex organs develop. They are vase-shaped archegonia (female organs), each containing an egg cell, and round antheridia (male organs), little bags in which swimming male cells are produced. The male cells swim by means of tiny beating hairs to the egg cells, which they fertilise. From the fertilised egg cell a new fern plant develops. At first it absorbs food from the little prothallus, but later it makes its own food by photosynthesis.

Club-mosses are quite different in appearance from the ferns. They have creeping, branching stems covered in whorls of tiny overlapping pointed leaves. Some grow hanging from tree branches. The spores are produced on special upright branches of different leaves, with spore cases on their undersurfaces.

Selaginella has made a big advance over the other club-mosses and ferns. It produces separate male and female spores. The small male spores are produced at the top of the cone, and the larger female ones at the base. The female spores remain on the plant until after fertilisation, rather like the egg cells of flowering plants. A male or female prothallus develops inside each spore. Then the male

Above left: The giant fronds of a tree fern in Portugal outlined against the sky.

Left: Young fronds of the royal fern uncurling in the spring.

Above right: The erect stems of the stag's-horn club-moss.

Right: The fertile spike of a giant horsetail. The surface of the stem is very hard.

cells swim to fertilise the egg cells in the female spores. The fertilised female spores are shed, together with a food supply in the form of the tiny female prothallus inside.

Horsetails are strange plants that have almost no leaves. They have existed on Earth for 325 million years. Their stems are green, and they photosynthetise. These stems are rough and ridged, and grow up from a creeping underground stem. At each joint of the stem is a sheath of scales—the remains of the leaves. A whorl of branches develops at each joint. The centre of the stem is often hollow, letting air down to the underground parts; this helps horsetails to live in very wet places.

The quillworts

Most ferns, club-mosses and horsetails grow best in damp places, but some members of one group of plants related to the club-mosses live submerged in lakes. This is the family of quillworts, which also have the more romantic name of Merlin's grass.

The common quillwort's roots are anchored in the bottom of a pond or lake. It has between twelve and twenty stiff pointed evergreen leaves.

131

Mosses and Liverworts

The mosses and their relatives the liverworts belong to the plant group called the Bryophyta. They are much smaller, simpler plants than the ferns, and are probably very similar to the first plants that ever colonised the land. Unlike the ferns and flowering plants, the bryophytes never developed water-proof skins. This means that they would quickly wither in dry surroundings; so they remain restricted to living in wet places.

The mosses have bodies clearly made up of stem and leaves, but they have no true roots. They are anchored to the ground by thread-like hairs called rhizoids, only a few cells long. Mosses have no real internal transport system for fluids, just a few central cells which are rather longer than the rest.

There are two main types of moss body. Upright mosses form tufts of erect shoots, and their stems are covered in whorls of leaves. Feather-mosses produce flattened, much-branched fronds close to the ground or to walls or tree trunks. Some develop attractive red and yellow autumn colours. Flat fork-mosses have leaves with little pockets on their under-surfaces for holding water. Some mosses, such as the hair mosses, have dark green shiny leaves, while others are pale green, such as the thread moss, or even translucent. Mosses such as cord moss form carpets; the screw moss often seen on walls and roofs grows in round hummocks.

These little plants are really the equivalent

Below: The spore capsules of moss on their long stalks.

of the fern's prothallus (see pages 130-131). At the tips of their branches sex organs develop, very similar to those of the ferns—archegonia and antheridia. After fertilisation, the egg cell develops into a spore case on a long stalk. This spore case is a surprisingly complicated structure. Both the case and the stalk are green, have stomata (pores) for air to enter, and can carry out photosynthesis to make their own food.

The spores are surrounded by sterile tissue, and conducting strands connect the spore case to the parent moss plant. The spore case is topped by a lid which falls off when the spores are ripe. Often there is a papery cap over the whole case. Beneath the lid the opening of the case is made up of a ring of teeth which open outwards in dry weather to let the spores blow away, and close again in wet weather. This prevents spores being shed in rainy weather, when they would have no chance of travelling far. The spores develop into little branching green threads, and new moss plants grow from buds on the threads.

The bog mosses, often called sphagnum

and *Lophocolea*, have definite leaves and stems, with the leaves arranged in two overlapping rows. All are anchored to the ground by tiny threads, each just a single cell.

A few liverworts, such as *Lunularia* and *Marchantia*, can reproduce asexually by producing tiny plantlets in special cups. The plantlets are scattered when they are splashed out of the cups by raindrops. In sexual reproduction, antheridia and archegonia are produced under flaps of tissue on the fronds, or on special shoots of leafy liverworts. The spore capsules are usually round and shiny black, splitting into four parts to release the spores. Tangled up among the spores are tiny threads called elaters, which twist sharply as they dry out, helping to flick out the spores. There are a few exceptions; *Riccia* produces spore cases tucked into the surface tissue of the frond, and *Marchantia* produces tall umbrella-shaped spore cases.

mosses, live in places where the soil is always full of water. They are the main vegetation of peat bogs. Each plant has a long main stem and bunches of side branches covered in tiny leaves. The leaves contain hollow cells with small holes in their surfaces which can hold many times their own weight of water, like a sponge. They grow in dense masses, and individual plants may be up to 300 millimetres (12 inches) long. The lower end dies, but does not decay because the acid water in which it lives preserves it. Instead, it forms peat. The spore case has a circular lid which blows off explosively, scattering the spores.

On boulders in mountainous and arctic regions grows a reddish-black moss called slit moss. It produces a spore case which opens by four slits, rather like a Chinese lantern.

The liverworts are much simpler plants than the mosses, and live in even wetter places, often partly submerged. Some, such as *Pellia* and *Marchantia*, are merely flat fronds while others, such as *Riccardia* and *Fossombronia*, have wavy edges. Some have midribs, others do not. The leafy liverworts, such as *Porella*

133

Fungi

The tasty field mushrooms, the green powdery bread mould and the blue veins of a good cheese may not appear to have much in common, but they are all fungi. The body of most fungi is made up of many fine threads called hyphae. Some hyphae are divided up into 'cells' by cross-walls; others are continuous tubes. A fungus does not have cells in the sense that other plants do. Each compartment of a hypha may contain many nuclei and a large central water-filled bag. The rest of the cell contents, including the nuclei, form a lining to this tube. Fungi have cell walls, but they contain no chlorophyll (green pigments), so cannot make their own food from carbon dioxide and water as green plants do. They can feed only on ready-made organic materials which they first digest externally with the aid of juices secreted by the hyphae, and then absorb.

Some fungi are parasites, obtaining their food directly from other living organisms. Others live on dead material such as leaves, the bodies of animals, leather or bread. They play an important part in breaking down dead organisms to return their mineral contents to the soil, so that the minerals can be taken up again by other plants.

Because they do not need light for photosynthesis fungi can live in dark places. Many of them live in the soil or under tree bark as fine networks of branching hyphae called mycelia. The familiar mushrooms and toadstools seen above ground are the fruiting bodies of these mycelia. Fungi have no obvious male and female plants, but there appear to be different strains of mycelia, and when two unlike strains meet they fuse and produce fruiting bodies.

A mushroom starts off as a knot of hyphae in the soil, which grows into a miniature mushroom. It then absorbs water, swells up and emerges into the air, opening out into an umbrella-shaped structure. On the underside of this umbrella are rows of ridges called gills. The tips of the hyphae forming the gills produce spores which fall off when ripe and blow away on the wind.

Another group of related fungi produces many pores (small holes) on the undersides instead of gills. Spores are formed inside the pores. This group includes the polypores, some of the commonest bracket fungi. Some fungi produce fruiting bodies once a year. Many of the bracket fungi last all year, and add more tissues to their fruiting bodies every year.

All these fungi belong to a group called Basidiomycetes (club fungi), which also

Field Mushroom   Destroying Angel   Red-Staining Inocybe

Fly Agaric   Tawny Funnel Cap

Common Stinkhorn   Horn of Plenty   Broad-Gilled Agaric

Shaggy Ink Cap

Rufous Milk Cap

Bitter Boletus

Hare's Ear

Graceful Brittle Cap

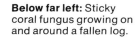

Below far left: Sticky coral fungus growing on and around a fallen log.

Below left: The mould *Penicillium notatum* (an ascomycete fungus) growing on the surface of agar, a kind of gelatin. It was from this kind of mould that the antibiotic penicillin was first made.

Below: Moulds growing on a rice pudding. Food that develops mould in this way should be thrown away.

includes the colourful jelly fungi, the brown earth-stars, the pretty branching coral fungi that grow up through leaf litter, and the rusts and smuts which attack crops.

The Ascomycetes (sac fungi) have a different kind of fruiting structure. The spores are produced inside a sac or bag called an ascus, and the asci themselves are grouped inside a fruiting body. The spores are shot out of the ascus by a build-up of pressure inside. Some varied shapes of fungi belong to this group —the morels, which look like round honeycombs on stalks, the up-curved cup fungi, some resembling crumpled bags, and the yeasts, so popular for their use in bread-making and brewing. They have some less popular relatives, however, such as the leaf-curl fungi, and *Aspergillus*, which often ruins bread and forms the white mould that attacks leather. Some species of *Aspergillus* cause lung and ear diseases in animals.

There are some even simpler fungi, called the Phycomycetes. Unlike the Ascomycetes and Basidiomycetes, their hyphae have no cross-walls. They produce no special fruiting bodies. The union of cells from different strains produces a fusion cell, which develops a tough coat and forms a spore. Such spores can survive severe conditions of drought and extremes of temperature. This group includes the pin moulds of bread and leather, and many fungi of stagnant water, whose white threads can often be seen on dead twigs and insects.

Some fungi are good to eat, but others are highly poisonous, and can kill you if you eat them. The best rule to follow is to eat only cultivated mushrooms, or fungi selected for you by a fungus expert. For more information about poisonous fungi see pages 144-145.

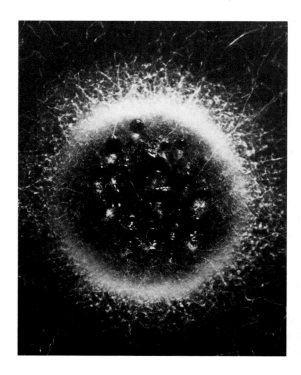

Algae

The simplest plants of all are the algae. One of the best places to see them is on a rocky shore at low tide, where the rocks and pools are festooned with red, green and brown seaweeds—all algae. They make their own food by photosynthesis, but they have pigments of various colours.

Algae range in size from single-celled forms such as the green powder, *Pleurococcus*, which you can see on walls and trees, to the giant kelps which can be more than 180 metres (600 feet) long. Many algae are known only by their Latin names.

Green algae appear in almost every drop of stagnant water, from the contents of water-butts to the water in a flower vase. The larger ones live in ponds and streams and in the rocky pools washed over by the sea. In the seashore pools the sea lettuce waves about like a sheet of bright green Cellophane, and feathery tufts of *Cladophora* hide under the seaweeds. Around the edge are the tubular fronds of *Enteromorpha,* a seaweed which some people use as food. Delicate green filaments grow in quiet fresh water—*Spirogyra* with its spiral green bands, and *Draparnaldia* with its clusters of finely-divided branches. Single-celled algae, some in ball-shaped colonies, bowl their way through the water. One single-celled marine alga, the sea bottle, has cells each the size of a marble.

The main occupants of rocky shores are the brown seaweeds. Layer upon layer of them coat the rocks, a slimy array of flat yellow-green to brown fronds. Many of the brown seaweeds are surprisingly complicated structures. They have flat, sticky organs called holdfasts for clinging to the rocks, and thick midribs containing conducting tubes almost as complex as those of the higher plants. They can grow wider and longer indefinitely.

Most brown seaweeds have many branches which let the waves rush between their fronds. Otherwise they would not survive, for they live in one of the most violent habitats on Earth, where waves incessantly pound the rocks. Many of these seaweeds have air bladders that help them float upright under water when the tide is in. At certain times of year the tips of the fronds become swollen with the reproductive structures—round hollows containing male and female sex organs. These sex cells are released into the sea, where they join together and are carried great distances. A few of the brown algae do not cling to the rocks, but float on the surface of the oceans. The Sargassum weed covers large areas of the Sargasso Sea in mid-Atlantic.

Adding a touch of colour to the rock pools,

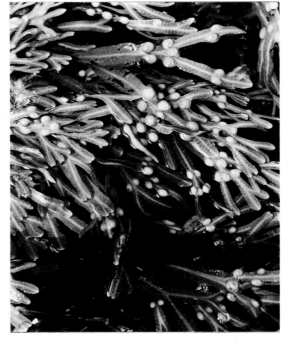

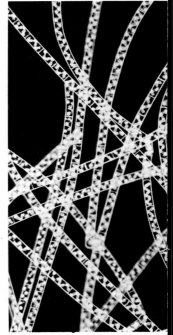

Above: Oarweeds, a form of brown algae, exposed at low water on a beach.

Above right: The lime-encrusted branches of coral-weed, which looks more like coral than seaweed.

Far left: The air-holding fronds of bladder-wrack, which help the plants to float upright when they are covered by the sea. The sex-cells are carried in the tips.

Left: This greatly enlarged photograph shows the microscopic freshwater alga *Spirogyra*.

but also growing deep out to sea, are the red seaweeds. A few form flat fronds like the brown algae, but most are made up of delicate branched thread-like filaments composed of chains of cells. An exception is *Delesseria*, which produces thin leaf-like fronds with wavy edges, of almost a fluorescent crimson. *Porphyra*, the purple laver, forms a flat wavy frond like a purple version of the sea lettuce, and is eaten in Wales as laverbread. Carragheen, or Irish moss, has a flat much-branched frond. It contains a jelly substance which people use in cooking. Many feathery tufted red seaweeds grow in rock pools. Carpeting wet shady rocks is a curious little red seaweed called coral-weed. It is encrusted with lime, so it looks more like a coral than a seaweed, especially when its lime has become bleached by the Sun. Very few red algae grow in fresh water. The red algae are much smaller than the browns, reaching at most a length of 1 metre (3 feet 3 inches).

Diatoms

Diatoms are tiny one-celled algae which occur in their millions in the surface waters of oceans and lakes. They form part of the mass of minute floating plants and animals called plankton. They are an important source of food for the animals of the plankton.

Diatoms have glassy shells in two overlapping halves, like the base and lid of a pill-box. Delicate strands of living matter emerge through tiny holes in the shell. If you look through a powerful microscope you can see that the shells have intricate patterns of grooves and dots, which make them look like sculptures in miniature.

Plant Specialisation

Just as humans specialise in the jobs they do in order to gain some advantage in life, so some plants have become specialists, which gives them a better chance of survival. Energy for making plants' body tissues comes from light, so light is their most important need for successful living.

A number of plants have developed ways of climbing high to reach the light without having to spend energy and materials to strengthen their stems. Some simply remain weak and become 'scramblers', leaning on any other plants that happen to be around. They usually have backward-pointing hooks or thorns which help them hang on to their supports. Many scramblers flourish in hedgerows—brambles, wild roses and goosegrass.

Plants such as runner beans, hops, vines and the tropical morning glory twine round other plants. Twiners are remarkable in that each species always climbs the same way, clockwise or anticlockwise. Some climb with their stems, while others—for example sweet peas—develop twining tendrils in place of some of their leaves.

Ivy goes a step further, anchoring itself to tree trunks and walls by thousands of tiny sucker-like roots, and the virginia creeper that covers some old houses has little branched suction pads. Bryony, which may straggle from tree to tree, has tendrils like coiled springs that stretch as the wind moves the trees. Lianas and honeysuckle often twine around each other for mutual support.

The banyan fig of tropical forests is even more remarkable. If a seed lands high in a forest tree it germinates, and the seedling sends down a long root to the ground, sometimes 60 metres (200 feet) below. There are many other plants which do not need to grow up to the light, but start off life growing on the branches and trunks of trees. Such plants are called epiphytes. They include many beautiful trailing orchids, succulent bromeliads, mosses and ferns.

Since more than 90 per cent of a plant's body is water, light is of no use unless the plant has a water supply. Very few plants are able to stand drying out. A few mosses on walls and rock surfaces can dry right out, then rapidly become green again after rain, but most mosses and liverworts are particularly vulnerable to drying out, and a few of them have developed structures for storing water. Sphagnum moss has water-storing cells over its surface, and the liverwort *Frullania* has a small pitcher at the base of each leaf. The cacti and succulents are the most efficient water storers. Most of their bulk is made up

Right: The prickly leaves of Mexican poppy, a tropical plant, help to keep it from being eaten by animals – an example of protective specialisation.

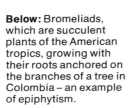

Below: Bromeliads, which are succulent plants of the American tropics, growing with their roots anchored on the branches of a tree in Colombia – an example of epiphytism.

Right: A baobab tree has an immensely thick trunk, in which water is stored. These trees grow in Africa, and this specimen was photographed in Kenya.

Below right: This species of palm tree growing in the Seychelles Islands has a system of prop roots which help it to stand in marshy ground.

of water-storage tissue, surrounded by a thick waterproof skin. The cactus body is ridged in such a way that it can expand rather like a concertina when there is plenty of water, and shrink again as it uses up its store.

Because they represent an important source of water, plants in dry areas are much sought after by animals. Plants have acquired a protective array of spines, thorns and other devices that kill their attackers or make themselves unpleasant to eat. The simplest defence is the bark of trees, a dead corky layer which deters many would-be parasites. Many shrubs, cacti and trees, such as hawthorns and acacias, have spines or thorns. Even leaves may be prickly, as anyone knows who has trodden on a thistle. Others develop stinging hairs, such as the stinging nettle and the cow-itch. Some have their stems covered with sticky glands, which prevent insects not useful for pollination from stealing the nectar. Many plants contain chemicals which give them an unpleasant taste or poison animals eating them. Perhaps the most unusual defence is that of the bull's horn acacia, which is protected by a private army of ants. It has hollow thorns in which live ants that swarm to the attack if any animal attempts to eat the acacia. A few plants rely on camouflage; the *Lithops* of African deserts is a succulent which looks like the stones among which it grows, and therefore is hard to see.

Parasitic Plants

Not all plants make their own food by photosynthesis. Some steal food direct from the living tissues of other plants and even animals, thereby saving themselves the need to develop green leaves or to grow in light places. Such a way of life, living at the expense of others, the hosts, is called parasitism, just as it is in animals (see pages 100-101).

Many of the lowliest members of the plant kingdom, the bacteria, cause diseases in humans, and in plants and animals too. They invade living cells and feed on their contents, and their waste products often produce unpleasant symptoms in their hosts. Usually it is in the parasite's interest not to kill its host, its source of food, but some fungi and bacteria are able to continue feeding on the dead tissues of their hosts, for example, on fallen tree trunks.

Fungi are the most numerous parasites. The mildews, moulds, smuts and rusts of crop plants are all parasitic fungi, carried from plant to plant as tiny spores on the wind. When a spore germinates it develops a network of thread-like filaments whose tips produce digestive juices. These juices dissolve the tissues of its host. Suckers on the filaments then absorb food from the living cells. Some fungi can even move from host to host through the soil. The potato blight, cause of the great Irish famine in the 1840s, produces swimming cells that travel in the soil water from one potato to another. The honey fungus, or bootlace fungus, produces honey-coloured toadstools on rotting tree stumps —and the fungus is responsible for the rotting of the tree in the first place. It spreads through woodlands as black strings of underground filaments looking like bootlaces. Even in water there is no escape from parasitic fungi. Some kinds grow over the gills of fish until they suffocate. There are also a few parasitic red seaweeds which live on other seaweeds. Some fungi which attack humans cause diseases such as athlete's foot.

A few flowering plants are able to live in dark corners of woodlands by living on the roots of other plants. Toothwort grows on elm and hazel roots underground, putting suckers into their tissues. It appears above ground only in spring, when it produces spikes of creamy tooth-shaped flowers. The leaves are hollow structures, almost colourless, serving no obvious purpose. The yellow bird's nest lives on decaying vegetation. In sunnier places broomrape grows on the roots of gorse, broom and clover, putting out flowering spikes with a few scale-like leaves.

Dodder is a parasite which taps its food

Above: Bracket fungi growing as parasites on the trunk of an oak.

Left: The scaly stems of broomrape, which is a parasite on plants bearing pods.

directly from the food-conducting tissues of its host. If a young seedling of dodder comes into contact with nettle, heather or clover plants, it sends out suckers into the stem to absorb the host's food materials. The tiny root withers, and the dodder twines round and round its host, sending out suckers at intervals. When its host flowers, the dodder absorbs some of the host's 'flowering signal' and produces clusters of pink flowers which form many tiny seeds.

Not all parasites are totally dependent upon their hosts. Eyebright is a normal-looking plant with green leaves, but its roots put out suckers to steal food from the roots of grasses. Many of the figworts also supplement their diet by drawing water and minerals from other plants.

Biggest flower

The greatest root parasite of all is the tropical plant rafflesia, which grows on the exposed roots and stems of trees in Malaysia. The only part to emerge is the flower, a gigantic scarlet bloom up to 1 metre (3 feet 3 inches) across and weighing 7 kilogrammes (15 lb). It is the largest single flower in the world.

This beautiful bloom smells like rotting meat, and attracts flies just as rotting meat does. Its local name, translated into English, is 'stinking corpse lily'.

Plant Communities

Almost all plants need sunlight, water, mineral salts and a good supply of air in order to make their own food. The closer together they grow, the more intense is the competition for these essentials of life. Even a single tree has its leaves arranged so that they do not overlap and shade each other. This arrangement forms a pattern called a leaf mosaic. The pattern of plants on a woodland floor reflects the amount of light getting through the leafy canopy above. There are patterns of plants in time as well as space: for instance, some woodland plants put out flowers and leaves early in spring before the trees come into leaf.

There is competition in the soil as well. Some plants have developed widely-spreading shallow roots while others penetrate deep into the soil. In this way the plants are not all trying to obtain water and minerals from the same layers of soil. This is particularly noticeable in deserts, where plants such as creosote bushes show surprisingly regular spacing. The distance between them is the area of soil the roots need to exploit to get enough water. In places where there is more rain the plants grow closer together.

In this struggle for survival some plants have developed individual ways of keeping other plants at a distance. The brittlebush is always surrounded by a ring of bare soil because its roots produce a poison that prevents the growth of other plants. Walnut seedlings protect themselves in the same way. Plants living underground have the same problems. Certain soil fungi produce substances which kill bacteria, so preventing them from competing for the dead organic matter upon which they feed. A number of these substances, such as streptomycin and penicillin, form very effective antibiotics for use by humans, too.

Some plants have evolved very successful special relationships with other plants. Many forest trees and members of the heather family are unable to grow unless certain fungi are present in the soil. These fungi grow over the surface of the roots, and sometimes even inside them, forming an arrangement called a mycorrhiza, which means fungus-root. The fungi apparently take over the function of the root hairs, drawing up water and minerals from the soil. Some of these fungi produce chemicals which join with certain minerals, making them more easily absorbed. In return the fungi get food made by the green leaves of the plant. As with animals, such mutually beneficial associations are called symbioses.

One of the commonest examples of symbiosis is provided by the lichens, the orange or

Above: Tree lungwort, a species of lichen, growing on a fallen birch tree. Lichen consists of a fungus and an alga living together.

Left: These coral-root orchids, growing in a Scottish pine forest, have a symbiotic relationship with a kind of fungus which grows over the surface of the roots.

Above: Young beech leaves in sunlight show the pattern known as leaf mosaic, in which the leaves of a tree are so arranged that they do not shade each other.

grey-green plants often seen as crusts on walls, roofs and rocks. A lichen is made up of a fungus and an alga. The main body of the lichen is made up of fungus filaments, scattered among which are algal cells. Alternatively, the algae may occur in distinct zones in the lichen body. The fungus provides the alga with protection from drying out, and produces acids which break down the rock to release minerals, which it then absorbs. The alga provides organic materials by photosynthesis, using water and mineral salts taken up by the fungus, and carbon dioxide from the fungus's respiration. During photosynthesis the alga produces oxygen, which is used by the fungus for respiration.

Lichens are extremely slow-growing, so they make very few demands on their surroundings. For this reason they can grow even in the most extreme environments, such as rock deserts and the edge of permanent snow and ice. They are the only plants able to break down bare rock. When they die they provide organic material for soil development.

Some plants use bacteria to help them take up nitrogen from the soil. Peas, beans and their relatives have lumps called nodules on their roots. These nodules contain certain bacteria which convert the nitrogen from the air trapped in the soil into organic nitrogen compounds that the plant can use. In return, the bacteria receive protection and organic substances made by the plant, which provide food for them.

Animal links

Certain plants form associations with animals. The bright colours of many corals are caused by algae living in their tissues. The algae use the phosphates and nitrogenous materials from the waste of the coral animals, and the animals benefit from the oxygen produced by photosynthesis in the algae.

143

Poisonous Plants

You have probably had painful experiences with stinging nettles, and been warned from a very early age that eating toadstools can kill you, but you may not realise just how many wild plants are poisonous. Not all are likely to cause death. Many are poisonous only if eaten in large quantities. They may in fact be beneficial to Man in providing valuable drugs when taken in the correct doses.

Toadstools are some of the most attractive-looking poisonous plants. Many poisonous fungi have brilliant colours, usually orange, yellow or red, which serve as a warning to would-be diners, and so protect the plant. The toadstool is the fruiting body of the fungus, and if it were eaten the spores of the fungus could not be dispersed. Any animal eating one of these fungi and feeling ill as a result would easily recognise it next time.

Not all poisonous fungi advertise their presence. The death cap or destroying angel is pure white, and some of the earthballs would pass unnoticed except to a trained eye. Some resemble only too closely the edible common field mushroom.

Among the most useful poisons for use as drugs are the alkaloids—chemical compounds probably occurring in 10 per cent of all plant species. These alkaloids include the heart drug digitalis, found in all parts of the foxglove; the many products of the opium poppy, including opium, morphia, codeine and papaverine; the deadly poison of the hemlock plant, colchicine, swallowed by the ancient Greek philosopher Socrates to commit suicide; hyoscyamine, from henbane, used as a mild anaesthetic, and as a poison by the infamous London murderer Dr Hawley Crippen in 1910; and atropine from the deadly nightshade, which is used to dilate the pupils of the eyes for medical examinations.

One of the most deadly alkaloid-producing plants is the thorn apple. Even a small bite of its beautiful white trumpet flowers could be fatal. In India it is used by thieves and assassins to knock out their victims. One of its relatives is thought to have been used by the priests of Delphi in ancient Greece to produce fits, said to be due to divine power. The attractive berries of the deadly nightshade are famous killers. In the Middle Ages an anaesthetic for surgical operations was brewed from deadly nightshade, opium, mandrake and hemlock. Alkaloids occur in many common plants such as daffodils and lilies.

Other types of drug also come from plants. For instance, cortisone is extracted from various trees. Cyanide occurs in some plant poisons, and plants such as rhubarb, which has poisonous leaves, contain burning acids. We use some plant poisons to poison other creatures. The insecticide derris is produced by the tropical derris plant, and pyrethrum comes from certain species of chrysanthemum. The poisonous leaves of delphiniums are never attacked by insects, though slugs seem unaffected by the poison. The effect of plant poisons seems to vary with the type of animal: for example, rabbits are not affected by deadly nightshade. Berries of the European shrub *Daphne mezereon* were once used to poison wolves and foxes, and the rat poison strychnine comes from the St Ignatius's bean.

Some plants are unpleasant even to touch. The painful stinging hairs of the nettle and its West Indian relative, the cow-itch, are well known, and North America has an even worse plant, poison ivy, which produces itching blisters some hours after the skin has brushed against it. Even smoke from burning poison ivy can bring up this rash, which may last for several days.

Many familiar plants, even buttercups, are poisonous. The seeds of laburnum and wisteria can be fatal to children. Many berries are poisonous, including the inconspicuous white berries of the mistletoe.

Not all plant poisons affect everyone all the time. The effect often depends on the person's state of health, size (small people need less poison to have an effect), inherited susceptibilities, age and even what else has been eaten. The ink cap fungus has very little effect unless alcohol is also taken in. Then it produces reddening of the skin and a rapid pulse rate, even if the alcohol is drunk some time after the ink cap has been eaten.

Right: The scarlet and white cap of the Fly Agaric fungus *(Amanita muscaria)* is a warning signal. It is dangerously poisonous when eaten.

Far right: A close-up of the beautiful flowers of the monkshood *(Aconitum anglicum)*. All parts of this plant are deadly poisonous.

Below: The giant milkweed produces a poisonous milky sap when the plant is damaged – you can see it oozing out of the damaged leaf in this photograph. A tiny amount can kill a person, and some African tribes used to poison the tips of their arrows with it.

Below right: The woody nightshade *(Solanum dulcamara)*, also called bittersweet, is a highly poisonous plant: neither the berries nor the leaves can be eaten.

Super poison

The tropical manchineel is probably the world's most poisonous tree. It is a member of the spurge family and grows, among other places, in Florida.

Its milky sap and its fruits, which look like yellowish-green crab apples, are deadly. Even dew or rain dripping from the leaves can carry enough poison to make your skin itch badly. A drop in the eye will make you temporarily blind. Smoke from burning the tree causes headaches.

Safety rule

There is only one safe rule when tasting plants you do not know: **when in doubt, DON'T!**

Insect-eating Plants

The idea of hunting for prey usually conjures up pictures of men and animals, such as lions and foxes, stalking other animals. In certain areas there are plants, too, which kill animals for food. Naturally they are not free to run after their prey, but they have developed subtle traps and individual ways of luring their quarry to its death.

There is a good reason why plants eat animals. The insect-eating plants grow in soils that are very poor in nutrients—food materials —and in particular nitrogen. These plants absorb the nitrogen they need from the bodies of the insects they trap.

One of the rarest carnivorous (meat-eating) plants is the Venus's fly trap, found only in marshy parts of North and South Carolina. It has a rosette of leaves whose upper parts are hinged at the middle. The blades of the leaves are reddish, and they glisten with droplets of liquid resembling a plate of juicy meat. On each blade are six hairs. When an insect touches one of these hairs it causes the leaf to snap shut. The spiky edges of the leaves interlock to prevent any escape. The leaf then secretes digestive juices that dissolve the tissues of the insect, which the plant then absorbs. Eventually the leaf opens again, and the undigested remains are exposed, to be washed away by the rain. Venus's fly traps usually feed on insects, but they can absorb creatures as large as baby frogs.

The sundews are insect-eating plants which are found in many parts of the world. They use quite a different technique. The leaves are covered with sticky red tentacles, each with a round glistening blob at the end which is covered with one of the strongest glues known to nature. Their glistening appearance ('sundew') attracts insects, which stick to the tentacles. The lightest touch on a tentacle causes all the other tentacles to curl over the victim, sticking to it and suffocating it by gumming up its breathing holes. The tentacles then produce digestive juices that dissolve the soft parts of the body, and the resulting 'insect soup' is absorbed by the leaves.

Some of the most elegant insect-eaters are the pitcher plants, whose special vase-shaped leaves may stand up on the ground, like bottles, or dangle on slender twining stems. These pitchers, some as much as 500 millimetres (20 inches) deep, are often very showy, coloured in reds, cream and greens, with crimson or purple spots that attract their prey. Just inside the rim of the pitcher the sweet-smelling nectar advertises its presence.

The bottom of the pitcher contains a soup of digestive juices. The inside of the pitcher is

Above: The vase-shaped leaf of a young pitcher plant. The pitcher has a leafy lid over the top to keep out the rain.

lined with downward-pointing hairs covered in slippery wax. As the insect alights to sip the nectar, it slides downwards on the waxy hairs until it falls into the digestive soup and drowns. Why it does not try to fly away is an unsolved mystery. It has been suggested that the nectar contains a narcotic which drugs the insect into a drowsy state, so it does not realise what is happening.

The butterwort has a rosette of yellowish-green leaves with slightly upturned margins. The whole surface is slimy and gland-covered. Once an insect alights it slides on the slippery surface, and its struggles stimulate the leaf edges to roll inwards to trap it.

A relative of the butterwort, the bladderwort, is a successful underwater hunter. The bladderwort is a water plant with finely divided feathery leaves which have many small oval bladders or bags like rows of tiny stomachs. Each of these bags is a trap for tiny water animals. When ready for action, each bladder is like a collapsed balloon before it is blown up. At the narrow mouth is a valve which lets water in but not out, and several long fine hairs protruding into the water. As soon as a passing water creature touches one of these hairs, the valve opens rapidly. The sudden inrush of water sucks the animal into the bladder, there to die and be digested.

Above left: The slippery, partly curled, leaves of a butterwort, ready for an insect to alight on them.

Above top: The open 'jaws' of a Venus's fly trap, ready for a victim. Note the spiny edges which lock together to complete the trap.

Above: A small gnat trapped by the sticky tentacles of a sundew leaf.

On the brink of death

Although insects and other small creatures fall victim to pitcher plants, one animal actually lives inside the pitcher, just clear of the deadly liquid.

This apparently foolhardy creature is a spider, *Misumenops nepenthecola*. It gets its living by grabbing the insects before they fall into the pitcher plant's poison vat, and eating them first.

Attracting Insects

The riot of beauty seen in flowers all over the world today is really not designed for our eyes at all. Flowers have developed their rich colours and varied shapes because of their need to attract insects. Many plants depend upon insects to carry their pollen from one flower to another, and the insects in their turn rely upon the plants for food. The flowers usually produce nectar—a sweet-scented sugary solution—which insects drink. Some insects also eat pollen. Pollen and nectar are to insects what meat and potatoes are to people—sources of protein and sugar.

In order to alert the insect to the presence of the nectar, flowers need to have large showy petals that can be seen from a considerable distance, possibly even before the insect can smell the nectar. The colour of the flower is a clue to the kind of creature pollinating it. Insects do not see colour in the same way as humans. They cannot see some of the colours we see, but they do see colours that are invisible to us. Bees, for example, cannot see red, but can see ultra-violet—a colour beyond our range. Many apparently plain flowers are found to have beautiful patterns on their petals when photographed in ultra-violet light.

The most attractive colour to a bee is a deep violet blue. Pink flowers are likely to be pollinated by butterflies, while flowers such as the moonflower which open at dusk and attract numerous night-flying moths are usually pale coloured, showing up in the fading

Left: The upper picture shows the flower of the evening primrose as we see it – a bloom apparently all one colour with very little pattern to it. The lower picture, taken by ultra-violet light, shows quite clearly the strong central pattern which a bee would see, a pattern to lead the insect to the pollen.

Left: Red and orange predominate in this bird-of-paradise flower from South America. The colours attract birds that pollinate the plant.

Right: The long, leafy spathe of a cuckoo-pint (lords-and-ladies) cut away to show the interior of the flower chamber, with pollen-carrying flies inside. Hairs trap the flies among the female flowers for a day or two until the anthers open to release pollen which falls on the flies. The hairs then shrivel and the flies escape to pollinate another cuckoo-pint.

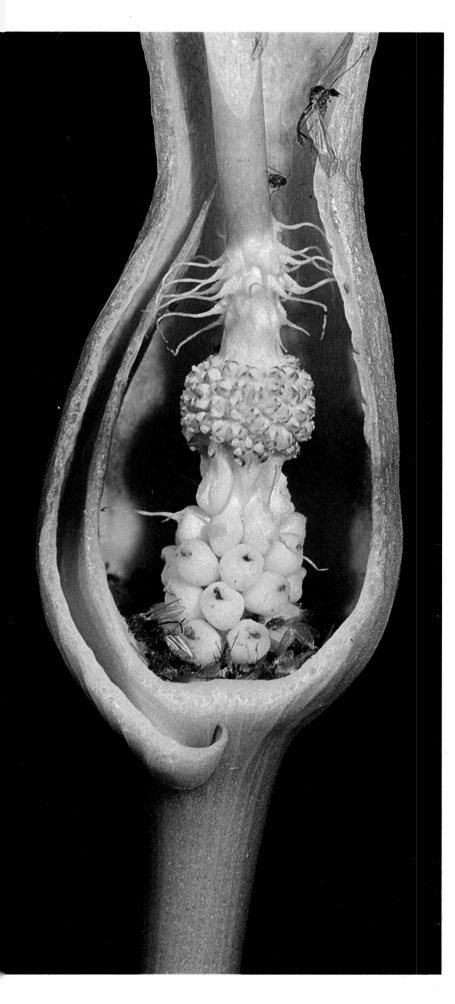

light. For humans the brightest colours are red and orange, and large flowers of these colours may be pollinated by birds or small mammals. Flowers pollinated at night by bats, which have poor sight, are often dark reddish brown in colour.

Once an insect alights on the flower, there are often still more aids to help it find the nectar. The centre of the flower may be a deeper or different colour from the rest, such as the yellow centres of daisies and the dark splashes at the base of pansy petals. There may be a series of guide lines or spots leading to the centre, as in foxgloves and orchids.

The shape of flowers may suit certain insects. Flowers such as the white deadnettle and snapdragon, pollinated by heavy insects —bees, for example—have landing platforms where the insects can alight. Butterflies have very long probosces, and they can feed from tubular flowers, but the opening of the flowers must be wide enough for the butterfly's wings. Narrow trumpet-shaped flowers are pollinated by moths, which have swept-back wings. Really flat-faced flowers such as speedwells (veronicas) are pollinated by hoverflies. Wide-mouthed flowers are also needed by humming-birds, and are produced by such showy plants as the hibiscus.

Other kinds of insects are enticed by different flowers. Flowers such as wild arum, polli-nated by wasps and flies, often have what we would consider unpleasant smells. Carrion plants have fleshy red-brown flowers that look like juicy plates of meat (and smell like them, too) which attract flies. Flowers pollinated by birds need to be robust, such as the banksia flowers of Australia (pollinated by honey-eaters) which look like brightly coloured brushes. Some African flowers open only when probed by a sunbird.

Flowers which use the wind to carry their pollen do not need such attraction. The wind-pollinated meadow rue is small and greenish, and the grasses have no petals at all. These plants need to be in a position to shake their pollen into the air currents, and many, such as oak and hazel, produce their pollen in long drooping catkins of male flowers which blow in the wind. The grasses have large anthers dangling on long slender filaments that pro-trude from the flower.

Wind-pollinated flowers have large feathery stigmas which catch the airborne pollen. The flowers must clear surrounding vegetation so that their pollen will not be trapped. For trees this is not a problem. Elms and ash trees produce their tufts of flowers before the leaves come out, but sycamores trail their catkins well below the leaves. Grasses and similar wind-pollinated herbs put up long flowering stalks.

149

HABITATS

The part of the Earth where a plant or an animal is normally found is called its habitat. You need to understand a habitat thoroughly in order to know exactly how each animal and plant in it lives. For over millions of years, as you can read on pages 170-171, all the living things we know today have evolved and adapted to suit the places where they live. Anything that did not adapt in this way was unable to survive, and became extinct.

The study of the relationship living things have with each other and with their habitat is known as ecology. A group of living things in a particular habitat is called a community, and community and habitat together form what scientists call an ecosystem. The total number of each species of plant or animal within an ecosystem is called a population.

To study an ecosystem, you need to know about the supply of energy and food in it and how that energy and food are used. Nearly all the energy comes from the Sun, and it is used by plants to manufacture their body materials from minerals in the soil and water—the process we call photosynthesis, described on pages 112-113. The food which plants produce is eaten by animals such as rabbits and deer, which are eaten in turn by carnivores such as foxes and wolves. When plants and animals die, insects and other small animals and bacteria help to break their bodies down into minerals which return to the soil, ready for reuse.

In the next ten pages we look at some different kinds of habitats and the way in which life goes on in them.

Barren-ground caribou in their summer habitat, the tundra of Alaska. These caribou get their name from the kind of territory where they are found. Here they graze on grass and the leaves of dwarf trees such as birch and willow. In the winter they migrate southwards to the evergreen forests.

Life Amid the Trees

Nearly one-third of the Earth's land surface is covered by forests and woodland of one kind or another. This is about half the amount of forest that used to exist before Man began clearing ground for his farms and towns. The clearance has made a very big change in the world's natural habitats, and besides the trees that have vanished, many animals and other plants have lost their homelands, too.

There are many different kinds of forests, and you can group them in several ways. The simplest way is according to the ecology of the area—climate, soil and the amount of rain that falls. These factors govern the kinds of plants that can grow and the kinds of animals able to thrive in a particular type of forest.

The most luxuriant forests are the tropical rain forests that grow near the Equator in Africa, Asia and South America, and in north-eastern Australia. The climate is warm and wet all the time. Most of the trees are ever-greens with broad leaves, and they grow very tall, forming a thick canopy high above the ground. This canopy keeps most of the sunlight from reaching the ground, and so there are only a few low-growing plants.

In these forests the trees flower and fruit at varying times of the year, so there is always a good supply of food for the vegetarian animal life. There are few large animals. At ground level live insects and the small mammals that feed on them, while in the treetops there are many brightly-coloured birds. Their gaudy colours, which look so obvious when you see them on their own, make them very hard to detect in the contrasting deep shadows and bright sunlight. Mammals include monkeys, fruit-eating bats, flying squirrels and other climbing species. The Australian rain forests contain the marsupial equivalent of some of these animals.

In swamps and near river banks, where the forest canopy is not so dense, the ground plants spread very thickly, forming the type of growth known as jungle. Jungle is very difficult to hack a way through. There are jungles in some of the other tropical forests where the climate provides definite wet and dry seasons. Many of the trees in these forests are deciduous, shedding their leaves for part of the year. They have forms of animal life similar to those of the rain forests.

There are two kinds of forests in the temperate regions of the world, deciduous and evergreen. The deciduous trees form a broad belt of woodlands across Europe, Asia and North America. The canopy is neither so high nor so dense as in the tropical forests, and there is a much richer variety of undergrowth

Above: A European beech forest in the autumn. Note the scanty ground vegetation under these trees.

Below: A typical coniferous forest. Again, there is comparatively little ground vegetation.

and smaller trees and bushes. Some trees, however, tend to check the growth of ground vegetation: the floors of beech forests, for example, are very bare. Deciduous forests are rich and green during the summer months, but bare in the winter. Other plant life tends to be seasonal: many small plants die back completely during winter and grow again in summer.

As a result the activity of animal life in these habitats varies, too. Many insects over-winter in the resting or pupal stage of their lives; some mammals, such as dormice, hibernate, while others, for example bears and

badgers, sleep for a large part of the winter months, waking from time to time to hunt for food. A great many of the birds of these forests fly south to warmer climates in winter.

The coniferous forests of the temperate regions are found mainly in coastal areas where the winters are mild and there is plenty of rain. Many regions have mixtures of evergreen and deciduous forest land, and the animal life varies accordingly. Like beech woods, coniferous forests contain much less undergrowth than the broad-leaved forests, and fewer species of animals live there. Further north in Europe, Asia and North America, and also in the mountain regions where the climate is cooler, lies the taiga, an exclusively evergreen forest land. Here the winters are cold, and life for animals is difficult. However, many species live there, including deer, bears and smaller creatures such as beavers, little rodents and hares. They in turn attract the predators—wolves, foxes and wild cats.

Above: The rich growth of a sub-tropical rain forest, with palm trees, screwpine, ferns and supplejack, a kind of liana.

Soil as a habitat

The soil provides a home for an enormous number of animals, many of them so small that you can barely see them with the naked eye, and some which you can only see with a microscope. Soil consists of rock broken down into a fine powder, mixed with the decayed remains of animal and vegetable matter—what gardeners call humus—and water.

Seeds, spores and roots in the soil provide food for some animal life, while many creatures depend on the decaying matter that eventually will form humus. Insects, centipedes and millipedes, mites, earthworms and the smaller, often parasitic, worms called nematodes all find a place to live in the soil.

Grasslands and Deserts

Woodlands are the parts of the world where life is at its most abundant. By contrast, grasslands and deserts are the wide open spaces of the Earth. Yet grasslands support a wide variety of living things, while deserts are comparatively bare—though few deserts are completely barren at all times.

Grasslands grow in regions of the world where there is not enough rain to provide the water needed for trees to grow. The main vegetation is grass of one kind or another, but other plants grow there as well. There are two main kinds of grassland. Savanna, or tropical grassland, has scattered trees, with a thicker growth of forest near rivers and lakes. The main savanna areas are in eastern Africa, the llanos of Brazil and the campos of Venezuela. The other kinds of grasslands, the temperate grasslands, go under various names in different parts of the world—the prairies of North America, the pampa of South America, the veld of South Africa, and the steppes of eastern Europe and the Soviet Union. Man has also created temperate grasslands by clearing forests in order to graze animals—for example in New Zealand.

Grass is a highly-nourishing food, and the grasslands support many kinds of grazing animals. In eastern Africa there are still large herds of antelopes and zebras, and also elephants. Giraffes feed on the leaves of the scattered trees. These herds, and those which graze the South African veld, were once much larger, but over-hunting has sadly reduced them. The herds move around the grazing areas in a vast circle with the seasons, so they never overcrop the grass—though in the confined areas of some of the wildlife reserves there is a danger that this may happen. The prairies of North America were formerly roamed by huge herds of bison—miscalled buffalo—but only a few thousand remain. The grasslands are now grazed by cattle and sheep, or are put under cultivation growing cereals—grasses of a different kind.

The big grazing herds are followed by predators—among them lions, hyaenas and jackals in Africa, jaguars in South America, and Tasmanian wolves (also called thylacines) in Australia, where they feed on kangaroos. The Tasmanian wolves are thought to have died out, though a few may still be living in Tasmania. There are also many kinds of birds, including the huge flightless birds—ostriches, rheas, emus—and birds of prey and carrion-eaters such as vultures.

There are two basic kinds of deserts: hot and cold. Hot deserts, where the Sun beats down mercilessly on rock and sand, have less

Right: A typical savanna scene in Tsavo National Park, Kenya, with scattered acacia trees and in the foreground a waterhole.

than 125 millimetres (5 inches) of rain a year, on average. This may even mean several years with no rain at all, followed by a downpour. Some deserts have four times as much rain, but because it falls in a few heavy storms and is quickly lost by running off the surface, soaking deeply into the ground, or by evaporation, it is of little value for plants and animals. On the other hand, some areas have only 250 millimetres (10 inches) of rain a year, but because it falls regularly and stays in the soil, forests can grow there.

A desert where nothing grows is rare. Most deserts have a few plants, described on pages 122-123. Where there are plants there is some form of animal life, mainly rodents, reptiles, birds and insects. Desert-living animals have adapted to the inhospitable conditions: many can live without drinking, getting all the moisture they need from the plants they eat, and conserving it so that their bodies lose as little as possible. They tend to be active at night, when the desert is at its coolest. Some animals survive the hottest time of year by means of aestivation, the 'summer sleep' (see pages 98-99).

Cold deserts are found in the lands bordering the Arctic Sea, where the climate is so cold that the subsoil is frozen all the year round. This region is known as the tundra. The whole continent of Antarctica is a vast cold desert, most of it buried under deep ice. Nothing can live there, except for a few plants, insects and spiders on the coast.

Left: The sparse vegetation found in a sandy desert. This is the Dasht-i-Kavir, in Iran, and a great deal of it is covered with salt.

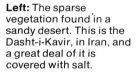

Mountain habitats
Life in mountain regions is always different from that in the country surrounding them because the climate is cooler. The temperature falls towards the mountain tops—between 5° and 10° C for every kilometre above sea level. As a result the conditions change as they do moving northwards towards the Arctic. A mountain may have its base in a tropical forest, with coniferous forest further up, followed by grassland and then a tundra-like terrain, and a peak covered with permanent snow and ice.

Animals of the mountains include many with thick coats against the cold, and nimble creatures such as goats and chamois which can move easily on steep slopes. These animals migrate up the mountain sides in warmer weather, and down again in wintertime.

Life Under Water

Nearly three-quarters of the Earth's surface is covered by water. A tiny part of this is fresh water in lakes, ponds and rivers. The rest is the salt water of the great seas and oceans. There is life in all these waters, except where the water is permanently frozen or has no oxygen in it.

Plants living in water can grow only at a depth where they can get enough sunlight. Below about 200 metres (650 feet) there is no light in the sea. In shallow water large plants can grow rooted in the mud, and often their tops emerge above the surface. On the seashore seaweeds are anchored to rocks, and in the tropics forests of mangrove trees cover mud-flats. Beds of reeds fringe lakes and rivers, and water lilies spread their broad leaves over still lakes and ponds.

In the deep sea all the plants have to be free floating. Almost without exception they are so minute that they can only be seen under a microscope. The cells of these tiny plants are so small that a hundred would fit on to one of the full-stops on this page. To feed on such minute plants, animals either have to be very small themselves, or else they need to have very fine natural sieves to strain the plant cells out of the water. Drifting freely in the sea with the tiny plants are myriads of little animals, which range in size from 1/10 millimetre to several centimetres. These drifting plants and animals are known as plankton. Biologists call the plants phytoplankton and the animals zooplankton.

The zooplankton include single-celled protozoans (see pages 88-89), a great variety of crustaceans, tiny jellyfish, larger jelly animals called salps, living in colonies, and a vast number of the larvae of bigger animals that, when adult, live in deep water or even on the sea bed. The zooplankton live by catching and eating the microscopic plant cells.

Some of the largest animals in the sea, such as the blue whale, basking and whale sharks and manta rays, live by feeding on the zooplankton. However, most of the larger, faster-swimming animals are active hunters, feeding on smaller fish and squid. So tuna, shark, giant squid, sperm and killer whales, seals and penguins correspond to the lions, tigers, jackals and eagles of life on land.

Most life in the sea is found in the upper layers, nearer the light and the plants, but some animals, fishes and other sea-creatures, are found even in the deepest waters.

Life originally evolved in the sea, but over millions of years descendants of some of the animals that adopted life on land have returned to the water, at least for part of their lives. Whales, seals, sea-cows, turtles, sea snakes and marine iguanas are all descended from land vertebrates, and they retain their land-based reliance on breathing air. Insects have not been at all successful at invading the sea, but they are important in fresh water, and many species spend the whole of their pre-adult lives in the water. Ponds, streams and lakes provide ideal habitats for the larvae of caddis-flies, dragonflies and mosquitos, as well as adult water beetles and water-boatmen. Some of the larvae have gills, so they can extract their oxygen from water. Others tap the air passages in the stems of water plants.

There is a larger variety of creatures living in fresh water than in the sea. This is because the chemicals in sea water are distributed fairly evenly, but the types and quantities of chemicals dissolved in lakes and rivers vary greatly. More kinds of large plants live in fresh water, and more plant material drops into it from dry land. The soil over which the water flows affects its contents. Hard water contains dissolved lime, whereas soft water contains no lime and is more acid. The kind of water has an effect on the animal life in it: for example, pond snails cannot survive in very soft water, because there is no lime available for them to make their shells.

Above: The whirlpool ramshorn snail lives in weedy pools and streams.

Above right: Tropical fish by a reef at Tobago Cays in the West Indies.

Far right: Salmon four days after hatching, with yolk sacs.

156

Seabed worms

Although there are no sea serpents or other marine creatures such as appear in legends, there are many strange creatures in the world's deep waters, and plenty more remain to be discovered.

One group of 'new' animals found in the late 1970s is a colony of giant worms, up to 3 metres (10 feet) long. They live 2.5 kilometres (1½ miles) down in the Pacific Ocean, west of Ecuador. At this point the water is warmed by hot springs welling out of the sea floor.

The heat from the springs provides energy—an alternative source to sunlight. Some bacteria are able to use this energy to manufacture their food by the process called chemosynthesis—a chemical version of photosynthesis. Small animals feed on the bacteria, and larger animals feed on the smaller ones, until big creatures like the worms and clams the size of dinner plates can find a living.

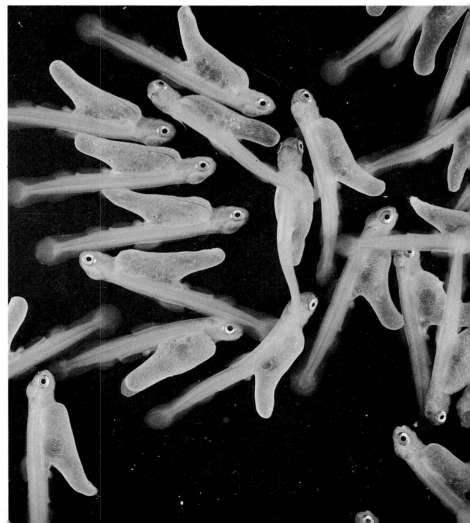

Food Chains and Webs

Out in the garden a cabbage plant grows. Its broad green leaves absorb the energy from sunlight and help the plant to make its own food and structural materials from water and minerals in the soil. Those leaves in turn are being eaten by the larvae of a number of insects. The larvae provide food for a sparrow—but this small bird is snatched up and devoured by a fierce sparrow-hawk, one of the birds of prey.

In this way the nourishment gained from the soil is finally consumed by the hawk. This is just one example of what biologists call a food chain, the who-eats-what line in nature. There are countless food chains in any habitat, and many of them are better called food webs, because the animals in them may each have more than one source of food. The sparrow, for example, eats a variety of insects, which in turn feed on many different kinds of plants. The sparrow may fall victim to a cat.

Food chains and webs are an essential part of what is called the balance of nature. This is the relationship of all the plants and animals in a particular ecosystem to one another. The balance of nature varies from one system to another. Over a long period of time the populations of different living things within a particular ecosystem stay constant, though there may be small variations from year to year. These variations cancel out in time.

A primary consumer, this red deer stag browses on grass, which biologists call a primary producer because the plant makes its food by means of photosynthesis.

To see how the balance of nature works, consider an area where there is plenty of grass and other vegetation which provides food for rabbits. The rabbits in turn are eaten by foxes and birds of prey. If there is a year of exceptionally good weather that encourages the growth of plants, there is even more food for the rabbits. As a result, more young rabbits survive, and the rabbit population grows. All this extra food helps the populations of foxes and birds of prey, and they too do well.

Now the balance of nature comes into play. The great increase in the number of rabbits helps to keep down the plants, so the area does not become overgrown; but the extra food supply of one year is probably not matched by next year's crop. The rabbits compete for food, and if there is not enough the weaker rabbits die. Meanwhile the large numbers of foxes and birds of prey are also killing off rabbits. Gradually the rabbit population falls to a level where the supply of plant food is just enough to support it. With fewer rabbits to eat, the number of foxes and birds of prey also drops, and all populations have returned to their normal level.

Biologists describe the plants of an ecosystem as primary producers, because they are the first producers of food. The plant-eating animals are the primary consumers, and the carnivores are the secondary consumers. The food chain may go beyond the secondary consumers: for example, in Africa lions are

secondary consumers, preying on antelopes (primary consumers); but the bodies of lions may in turn be eaten by vultures.

In normal conditions the operation of a food chain helps to keep a population of animals strong and healthy. As a general rule it is the weaker antelopes which fall victim to the lions. The stronger ones survive and breed. In the same way only the fittest lions can catch antelopes: the weakly ones and the old and sick cannot get food, so they die. Diseases and parasites also help to kill off the less fit members of a community. The stronger ones are less affected, so they survive.

The balance of nature can be upset by outside influences. Sometimes these outside influences are caused by natural events, such as forest fires and floods, or changes in climate, but most things that upset the balance are caused by the actions of Man. If he kills off too many predators then the number of primary consumers, the plant eaters, grows beyond what the supply of plant foods can bear, until starvation reduces the numbers again to a balanced level.

Another step in the food chain: a zebra, which is a primary consumer because it feeds on plants, is killed and eaten by a lioness, which is a secondary consumer.

The lemming story

Nature has its own methods of keeping the balance of animal life constant. In Lapland, the northern part of Norway, Sweden and Finland, there lives a little rodent called the lemming. Every few years the food supply of these animals becomes particularly plentiful, and the lemmings breed more freely. In no time at all there is a population explosion, and there are far too many lemmings for the area.

When that happens the lemmings, which are normally shy, night-working creatures, come out into the open in their thousands, and begin a mass migration. Once started on their journey, the lemmings seem unable to stop, and many eventually reach the sea or other water and drown.

Animal Homes

For many animals the whole of their habitat is their home. The great grazing herds of the African grasslands, the bison that once roamed the American prairie in their millions, the kangaroos of Australia all move from place to place, never staying long in any one spot. For many other animals one particular part of their habitat is a base, a home much as a house is for people. Some of these homes last only for a few days or weeks. Others last the lifetime of the animal, or even for many generations. Animals include some amazingly skilled craftsmen and architects—though this skill is inherited rather than learned.

Many of the invertebrates make homes for themselves. You can read on pages 80-81 how the social insects—bees, wasps, ants and termites—make elaborate homes, some of which last for many years. Their relatives, the solitary bees and wasps, also make homes in which to rear their families. For example, female digger wasps make burrows in the ground or in decayed wood. Each burrow is provisioned with live food—a caterpillar, fly, cicada or other suitable insect—which the wasp has paralysed with her sting. On or near this living larder the wasp lays an egg. Some species inspect their burrows daily, and if the egg has hatched the mother brings more supplies of food. Potter wasps follow a similar routine, but build their nests from lumps of clay. Mason bees also build nests, using sand or rock dust cemented together with saliva. Different species of solitary bees and wasps each make their own type of nest.

A more fragile home is constructed by the

Above: A harvest mouse peeps out from its nest. The mouse begins by interlacing the leaves of stalks which are standing together, then builds up the rest of the nest with blades of grass to form a firm ball.

Left: Young prairie falcons huddle in a rocky niche. Falcons do not build nests, but often lay their eggs on cliff ledges.

Right: A prairie dog at the entrance to its burrow. One member of each prairie dog group is usually on sentry duty.

larvae of froghoppers, which are among the true bugs. Each larva hides inside a little ball of froth, to which our ancestors, who could only guess what it was, gave the name of cuckoo-spit. It is often seen on plants.

Of the vertebrates, the birds are the most familiar home-makers. Their nests are many and varied in design. Some birds do not trouble to build a nest: many gulls, for instance, lay their eggs in a hollow on the ground, but most birds make some kind of safe home for their eggs and the nestlings. A great many birds build cup-shaped nests using a variety of materials—twigs, hair, feathers, bits of grass, straw and spiders' webs among other substances. The larger birds use more substantial materials: eagles use quite heavy branches and sticks, and so do storks. Some birds like a roof over their heads: wrens make spherical nests, while woodpeckers hack holes in the trunks of trees. Weaverbirds hang their nests from the ends of branches.

Swifts and swallows are among the birds which make their nests from mud and clay, rather like the potter wasps. Perhaps the most remarkable nests are those of the tailor birds, which sew the edges of large leaves together to make a bag, using their beaks as needles.

Most animals make their homes to suit their environment. One animal, however, changes its environment to make its home. It is the beaver. If conditions are right, the beaver makes its home in a hole dug into the bank of a river, with its entrance under water. The lodge, as it is known, is sometimes extended above ground level by piling twigs and mud on top. If the river is too shallow, the beavers —for several animals work together as a team—set about damming the river to make a

pond, in which they can swim and dive, and also construct their lodge. They cut down trees, gnawing them through with their sharp, powerful front teeth, then cut them up and drag the branches and the trunk down to the river. If there is no other way of getting a tree to the water the beavers dig a canal to float it.

Some beaver dams are huge: the largest known, on the Jefferson River in Montana, is 700 metres (760 yards) long. Dams of this kind are probably several hundred years old. The beavers constantly maintain their dam, and regulate the water-level in the artificial lake by raising or lowering the height of the dam.

Numerous other animals make underground homes. The biggest are probably those of badgers, which are called sets. A well-established set may be about 5 metres (16 feet) deep and contain a maze of passages connecting sleeping chambers and several entrances. Rabbits and foxes also burrow underground, though a fox often takes over part of a badger's set, acquiring a ready-made home.

Below: Low water in a beaver pool has exposed this beaver lodge – a large, untidy-looking mound of sticks and mud, with entrances under water.

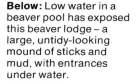

Animal towns

Prairie dogs, ground-living members of the squirrel family, live in such large communities in North America that people call them 'towns'. The animals make burrows with separate underground dens for up to 15 individuals, and several exits. Groups of these burrows may extend over an area of several square kilometres.

Similar but smaller communities are made by the marmots of Europe and Asia and the woodchucks of North America. These animals and the prairie dogs are all rodents, but one species of mongoose, the carnivorous meerkat of South Africa, makes a similar kind of home.

PREHISTORIC LIFE

We share our Earth with well over a million other kinds of animals, and many thousand of different plants, yet all these living things came from a few tiny specks of life that first appeared thousands of millions of years ago.

Each of the first living things was made up of a single cell. These organisms simply multiplied by splitting themselves in two. Sometimes new cells stuck to old ones, and built a larger organism. Some of these organisms had specialised groups of cells performing particular tasks—finding food, for instance. Gradually organisms appeared which were built in ways that fitted them for different kinds of life.

Early animals included sponges, sea anemones, jellyfish, worms and the jointed-legged ancestors of insects, crabs and spiders. All these early animals lived in the sea, and all lacked a backbone, but in time a worm-like organism gave rise to fishes—the first backboned animals. From fishes came amphibians resembling modern salamanders, frogs and toads. Amphibians gave rise to reptiles. Reptiles in their turn gave rise to birds and mammals.

This chapter shows how we know that life evolved. It explains how we discovered that countless prehistoric plants and animals died out. Here, too, we glimpse some of the astonishing plants and animals that lived long, long ago.

The Grand Canyon of the Colorado River in Arizona has been cut 1,900 metres (6,250 feet) through layer upon layer of rock. In the upper layers, which are sedimentary (deposited) rocks, is an amazing record of fossil life, from primitive algae at the bottom to dinosaurs and early mammals in higher strata.

The Fossil Record

More than 2,000 years ago Greek scholars noticed what seemed to be stone sea shells stuck in rocks. They rightly guessed that these were the remains of sea creatures that had died long before. Somehow the shells had turned to stone, and the sea bed in which they were buried had been forced up and turned into dry land.

Shells like these are now called fossils, from an old Latin word that means dug up. In the past 200 years geologists studying rocks have dug up and chipped out the fossils of numerous different kinds of plants and animals. We now know that many of these organisms died out millions of years ago, and therefore fossils are our most valuable clues to past life on Earth.

Fossils are formed in several ways. Most begin as the bodies of dead plants or animals that sink to the bed of a lake or the sea. Usually the soft parts soon decay, leaving only a skeleton of hard fibres or bones and teeth. Mud and sand pile up above the skeleton. In time the overlying layers become so thick and heavy that the soft soil around the skeleton is compressed and hardened into solid rock.

Meanwhile water seeping downwards deposits minerals in tiny holes within the skeleton. This makes the skeleton harder and less likely to be crushed by the surrounding rocks. Sometimes minerals replace all of the original substance. When this happens we say the plant or animal has been petrified—that is, turned to stone.

Sometimes water dissolves a shell or bone without replacing it with minerals. This leaves a hollow in the rock. If its shape matches the vanished shell or bone, this hollow is a fossil, too. Hollows in the form of missing fossils are called fossil moulds. In time movements of the Earth's crust may bend and buckle seabed rocks, forcing them above the sea. Later on rain, sun and frost wear away the surface rock, laying bare the fossils underneath.

People often find fossils exposed in rock layers in cliffs, mines and quarries. Usually the lowest rock layers are the oldest and the highest layers are the youngest. Experts use fossils to determine the age of the rock layers in which they are found.

Digging fossils out of solid rock without breaking them is difficult. Extracting a fossil dinosaur may mean first blasting or bulldozing tonnes of rock. Then the fossil-hunters must free the brittle bones with care. They chip away with chisels, gouge with awls, dig with trowels, and sweep loose soil away with wire brushes. When they have freed a large fossil

164

bone they paint it with preservative. Then they protect it with bandages and splints to prevent the bone from breaking.

Inside a museum, experts clean all the fossil bones belonging to one skeleton and try to reassemble them. They use as guides photographs taken to show how the bones were grouped before they were dug up. A museum expert may be able to rebuild a big fossil skeleton with the help of metal rods to support the bones. Careful study of these bones shows him how and where the muscles fitted and how the creature used to stand and walk. This knowledge helps an expert to make a life-like model of the fossil animal.

Careful study of its fossil remains can also reveal a great deal about how a prehistoric creature lived. The size and shape of bones and joints, with their relationship to one another, may show how fast it ran and walked, and if it was warm-blooded like a modern mammal, or cold-blooded like a snake. Claws and teeth help to show whether the animal used to eat plants or kill other animals for food.

Above: Students in the Fossil Gallery of the Natural History Museum in London sketching some of the reconstructed fossil skeletons on show there. The left-hand skeleton is that of an *Iguanodon,* and on the right of the picture is the skeleton of a *Diplodocus.*

Above right: The fossil remains of a trilobite, a sea animal which lived between 225 million and 600 million years ago. There were more than 4,000 species of trilobites.

Left: The skeleton of an *Ichthyosaurus,* a fish-like reptile, in the rock where it was discovered.

Preserved in ice

Fossils take many forms. For instance, the carbon in a plant leaf may survive as a delicate film on a rock layer. An insect may be preserved in amber—a hard yellow substance that was once sticky resin oozing from a tree.

Peat, tar and ice have amazingly preserved the bodies of certain beasts. People have even discovered woolly mammoths, complete with flesh and fur, embedded in ice in cold Siberia.

Such unchanged bodies are not strictly speaking fossils, even though they died so long ago.

165

The Great Reptiles

One day in 1822 an Englishwoman out for a country stroll noticed strange fossil teeth among a pile of roadside stones. Her husband, Dr Gideon Mantell, had never seen such fossils. They looked like those of an iguana lizard, but were much larger. He named their vanished owner *Iguanodon*, which means iguana tooth. Soon people were finding the fossil bones of many more unusual prehistoric beasts, seemingly giant reptiles. In 1841 a British scientist, Richard Owen, named the whole group dinosaurs, from two Greek words meaning terrible lizards.

We now think that dinosaurs evolved about 200,000,000 years ago from a small relation of the crocodiles. Because most of the continents were then joined together as one landmass, dinosaurs could and did spread around the world. People have found their fossil bones in every continent except Antarctica. For 140,000,000 years these giants were masters of the land.

Two large groups of dinosaurs emerged. One group had hip bones rather like those of a lizard. The other group's hip bones were rather like those of a bird. Within each group a great variety of creatures evolved.

Lizard-hipped dinosaurs came in two main types: sauropods and theropods. Sauropods, or lizard-footed dinosaurs, were the largest animals that ever lived on land. They had a huge barrel-shaped body, a long neck and tail, and legs like short, strong pillars. The heaviest kind was *Brachiosaurus*. Some specimens weighed as much as 20 elephants. The longest sauropod was *Diplodocus*, which measured up to 27 metres (90 feet). Despite their huge size, these dinosaurs were harmless plant-eaters. They used their long necks to crop leaves from trees.

The theropods, or beast-footed dinosaurs, make up the second subdivision of the lizard-hipped group. Some theropods were the most terrifying dinosaurs of all. Beasts such as *Tyrannosaurus* stood as tall as a giraffe. They walked and ran on huge, powerful hind legs. Teeth like daggers and claws as long as carving knives helped *Tyrannosaurus* to kill and eat giant sauropods.

The theropods included smaller hunters, too. Some were no bigger than a hen, and no doubt scampered after small prey such as lizards and large insects. Certain scientists believe that one species of these tiny lizard-hipped dinosaurs developed feathers and became the ancestors of the birds.

Meanwhile the bird-hipped dinosaurs—the second major group—were also giving rise to a variety of creatures. Unlike some of the

Pteranodon

Triceratops

Stegosaurus

Iguanodon

Diplodocus

lizard-hipped dinosaurs, all bird-hipped dinosaurs were harmless plant-eaters. As in the first group, some walked on two legs, some on four. Two-legged bird-hipped dinosaurs included *Iguanodon*. If *Iguanodon* were still alive it would be tall enough to peer through the upstairs windows of a house. *Iguanodon* was as heavy as an elephant, but spiky thumbs were its only useful defence.

Similarly built but stranger in appearance were the hadrosaurs, or big lizards. Like *Iguanodon* they measured about 9 metres (30 feet) long. They had flattened, duck-like jaws containing up to 2,000 teeth, designed to grind tough leaves to pulp. Differently-shaped bony crests sprouted from the heads of the several dozen different kinds of hadrosaurs.

Later came the ankylosaurs, low-slung monsters with armour-plated backs. A flesh-eating dinosaur could kill an ankylosaur only by overturning it and tearing at the soft skin of its belly. Some other four-legged dinosaurs in the bird-hipped group would have made even more difficult targets. Called horned dinosaurs, they resembled giant rhinoceroses. Spikes jutted forward from their faces and their necks were guarded by a massive bony frill. *Triceratops*, the largest horned dinosaur, weighed more than a bull elephant.

Dinosaurs were still plentiful 70,000,000 years ago. Then they all became extinct. No one knows the precise reason, but it seems likely that the climate grew too cold for them to survive. Small cold-blooded animals can escape the frost by crawling into holes, but most dinosaurs were far too large to do that. Even if the dinosaurs were not cold-blooded, as some scientists suspect, they certainly lacked fur to retain their body heat. Either way these prehistoric giants would have perished in the cold.

The original boneheads

Some of the oddest dinosaurs were a group of comparatively small animals which must surely be the original boneheads. There were several species, but the one with the thickest skull was *Pachycephalosaurus* —which means thick-headed reptile.

This beast had a skull nearly 25 times as thick as that of a human being. It was reinforced with sharp knobs on the back. Scientists think the males of these dinosaurs used their hard heads for duelling, just as present-day rams do.

167

Prehistoric Plants

In 1977 American scientists found tiny fossil plants in rocks about 3,400,000,000 years old. These plants were blue-green algae, related to the slimy blue-green growths you sometimes see growing on the sides of fish-tanks.

Like all green plants, these ancient algae used sunlight to obtain the energy for making food from the substances around them. They lived in water which supported them and held the chemicals they needed for their food. The first single-celled plants gave rise to more advanced kinds of plants made up of cells that clumped together. These algae came to include the familiar seaweeds. By 700,000,000 years ago simple water plants were being eaten by the first, water-living animals.

There were still no plants capable of life on land. This is easy to explain. Air is too thin and dry to support the soft body of an alga and to keep it damp. A seaweed stranded on a beach just shrivels up and dies.

In time, however, a scum of algae grew on the damp edges of rivers, lakes and seas. By 400,000,000 years ago these lowly plants were evolving into plants better suited to life on land. Land plants have roots which draw up minerals and water. Their leaves trap sunlight, and a stiff stem supports the leaves and carries minerals and water to them. Their waxy surface prevents them from drying up.

Early land plants lacked some of these

Above: A prehistoric forest probably looked very much like this. The artist has based this view on the kinds of plants of which we have found fossils. They include cycads on the right, with giant horsetails on the left and club-mosses in the foreground.

Left: The fossilised remains of a fern, which have been perfectly preserved in the rock.

168

spores produced new plants only if they fell on damp soil, so as the swamps dried out the coal forests died off.

Meanwhile there arose some plants that could cope with drier conditions. These plants produced two kinds of spores, one large and one small. At first the large spores stayed on the plants. Protecting layers grew around these spores, keeping them moist and nourished. The small spores blew away as pollen. When a pollen grain reached and fertilised a large spore the two combined spores became a seed. The seed then ripened and fell to the ground. If the ground was dry the seed's outer coat could keep it moist for months. When rain fell the outer coat burst open and the seed began to grow. This meant that seed-bearing trees could grow on land that was too dry for the trees bearing unprotected spores.

Early seed-bearers included the ancestors of the conifers—trees that bear their seeds between the scales inside a cone. Conifer seeds are not completely shut in and protected. Botanists call such plants gymnosperms, a word meaning naked seeds. By 100,000,000 years ago a new group of seed-bearing plants had developed. These had seeds completely surrounded and protected by a fruit. Botanists call them angiosperms, a word meaning capsule seeds. We call them flowering plants. For more information about plant life today see the chapter on *The Plant Kingdom*, which begins on pages 108-109.

Below: The horsetail of today is very similar in general appearance to the plants of prehistoric times, but those early horsetails grew the size of trees.

features. For instance, little *Cooksonia* was just a cluster of stems, but there were also club-mosses, rather like those still alive today. From other early land plants came the familiar ferns and horsetails.

Living club-mosses are low enough to tread underfoot, and most horsetails grow only waist high, but about 300,000,000 years ago there were club-mosses and horsetails that grew in the form of trees. Huge forests of these giants sprouted in warm, low-lying swamps in North America and Europe. When the trees died they fell into the swamps. Their dead remains in time formed a thick, peaty layer. Then the sea level rose. The sea flooded the swamp forests, covering up the peat with sand and mud. Later the sea level fell and a new forest grew up on the sand and mud. These changes happened several times. They built thick layers of peat, sand and mud. As the layers multiplied, their weight squashed the peat and gradually changed it into coal—which is why we call coal a fossil fuel.

By 225,000,000 years ago lands were getting drier. This made conditions difficult for the coal-forest trees because they grew from tiny unprotected particles called spores. These

Evolution

Nowadays most thinking people believe that one kind of plant or animal slowly changes over the years, giving rise to another. We now accept that living things have gradually evolved to their present state, but a century ago few people though that this was so. Most Christians, Jews and Muslims believed the Bible story which seemed to say that God had made every kind of living thing all at once, and that none had ever altered. People explained extinct fossil animals as species killed off by Noah's Flood.

By the early 1800s new ideas paved the way for belief in evolution. To begin with, the Swedish naturalist Carolus Linnaeus (1707-1778) gave people a clearer understanding of living things. He grouped and named them according to the likenesses he saw. For instance, he grouped together birds with similar bills and claws.

Between 1830 and 1833 the British geologist Sir Charles Lyell proved that the Earth was extremely ancient. This meant that there had been plenty of time for evolution to have taken place. Previously, people had tried to calculate the age of the Earth from the Bible; James Ussher, Archbishop of Armagh, reckoned that the creation occurred in 4004 BC.

Later in the 1800s the modern theory of evolution sprang separately to the minds of two British naturalists. The idea came first to Charles Darwin (1809-1882). In 1831 he set out as naturalist aboard *HMS Beagle* on an almost five-year voyage around the world. In South America Darwin found extinct fossil beasts that looked much like some living species. Darwin suspected that the extinct animals included the ancestors of living animals. On the Galápagos Islands, which lie in the Pacific Ocean more than 900 kilometres (550 miles) from the coast of Ecuador, he found finches and giant tortoises unlike any on the mainland. Darwin believed they had evolved there on the islands from ancestors that had somehow arrived from South America.

Why and how did evolution happen? Darwin knew that each species produces individuals that are slightly different from one another. He knew that each individual tends to hand on its special features to its offspring. By breeding from individuals selected for their special features, Man had produced different breeds of dogs, cats and cattle useful to him as domesticated animals.

Darwin reasoned that, in the wild, Nature did the selecting. He believed that Nature favoured those individuals and their offspring best adapted to survive in their surroundings by finding food, escaping enemies and with-

Charles Darwin first set out the modern theory of evolution. His work aroused a storm among people who thought that he was contradicting the Bible and therefore the word of God.

The evolution of the horse as we have learned it from fossils. The horse on the left, *Eohippus* (which means 'dawn horse') was the size of a fox. It lived 55,000,000 years ago.

standing heat or cold. If their surroundings gradually altered, natural selection might favour individuals with rather different characters. Thus over millions of years the struggle for survival killed off some species and created others.

Unknown to Darwin, another British naturalist separately reached the same belief in evolution by natural selection. That man was Alfred Russel Wallace (1823-1913). In 1858 Wallace wrote to Darwin about what he had discovered. Darwin told scientists about his own and Wallace's discoveries. Then in 1859 Darwin set forth his theory publicly in his book *On the Origin of Species by means of*

Eohippus

Mesohippus

Merychippus

170

Natural Selection. This famous work changed the way in which people looked at past life.

Darwin's book left unanswered some important questions, however. Why were individual plants or animals bigger, smaller or different in other ways from their brothers and sisters? Why was each one not exactly like its mother or father, or a mixture of the two? The Austrian monk Gregor Mendel (1822-1884) discovered that the answer lay in the hereditary factors passed on by both parents according to certain patterns. Scientists now call these factors genes (see pages 16-17).

A contemporary painting of Alfred Russel Wallace. His name is still commemorated in Wallace's Line, an imaginary line which divides the animal life of Asia from the very different animal life of the Australian region.

Lamarckism

The French biologist the Chevalier de Lamarck (1744-1829) worked out a theory of evolution before Darwin. Like Darwin, Lamarck thought all living things could have evolved from a common ancestor but, unlike Darwin, he argued that an animal could pass on to its young changes to its body organs that were brought about by use or disuse.

According to this way of thinking, that long-necked mammal the giraffe came from generations of animals that stretched their necks to reach the leaves on trees. Each generation supposedly had a neck a little longer than the one before.

In the early 1900s an Austrian scientist experimented with midwife toads and salamanders. He claimed he had proved that animals could inherit bodily changes produced in their parents, but tests by other scientists showed that cheating had taken place in the experiments. Scientists now reject Lamarckism—the belief that evolution can happen by the inheritance of acquired characters.

Pliohippus   Equus

The Evolution of Man

Long ago Man's ancestors were small, sharp-snouted, furry, long-tailed beasts. These tiny animals climbed trees after insect prey, and resembled modern tree shrews. They eventually gave rise to the primates—the group of mammals that includes the monkeys, apes and Man. Early primates were suited for life among the trees. They had toes and fingers enabling them to grip the branches; their big, forward-facing eyes helped them to judge distance when they leaped from branch to branch, and their other special features included a small nose and a large brain.

Fossil finds from Egypt show that the ancestors of Old World monkeys, apes and Man had evolved by 35,000,000 years ago. A direct ancestor of Man may have been the ape-like creature *Ramapithecus*, which lived 10,000,000 years ago. In 1960 the British anthropologist Louis Leakey found its fossil bones on a farm in Kenya. Its fossils have also turned up in places as far apart as Hungary, India and China. Teeth more like those of Man than modern apes helped *Ramapithecus* to eat many kinds of food. Perhaps its unfussy eating habits helped this creature to spread so far around the world.

Early primates went around on all fours and used their arms to grasp branches, but by 5,000,000 years ago *Ramapithecus* may have had a descendant that walked erect. Scientists call such creatures australopithecines (southern apes). Instead of living in the trees, these primates probably spent their time in the

The tree-shrew (above) is similar to the little, long-tailed animals that were the ancestors of Man and the apes. The tarsier (right) is a survival of an early stage of primate evolution.

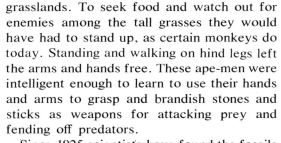

Below: An artist's reconstruction of *Ramapithecus*, an apelike creature which lived 10,000,000 years ago. It was more like an ape than a man, but its teeth were similar to those in a man's jaw.

grasslands. To seek food and watch out for enemies among the tall grasses they would have had to stand up, as certain monkeys do today. Standing and walking on hind legs left the arms and hands free. These ape-men were intelligent enough to learn to use their hands and arms to grasp and brandish stones and sticks as weapons for attacking prey and fending off predators.

Since 1925 scientists have found the fossils of the ape-man's bones in several parts of Africa. Some of the most striking finds came from Olduvai Gorge in Tanzania. In and after the 1950s Louis Leakey and his wife Mary between them found what scientists now believe to be the remains of two kinds of australopithecine. One was smaller than the other, but had a bigger brain. Close by this smaller creature's fossils lay chipped pebbles. This discovery showed that some ape-men were making tools nearly 2,000,000 years ago.

In the 1970s Louis's son Richard and other scientists made even more astonishing discoveries. Near Lake Turkana (once called Lake Rudolf) in northern Kenya Richard discovered the early fossil bones and stone tools of *Homo erectus* (Upright Man). Upright Man was taller than the southern apes and stood up straighter. He also had a bigger brain and a better grip. He made superior stone tools and knew the use of fire. Yet the fossil finds from Africa suggest he may have lived as long ago as the australopithecines, so scientists are not sure if these creatures could have been his ancestors.

Upright Man's ability to hunt and keep his body warm helped him to spread through Asia and Europe. Fossil finds show that by 300,000 years ago Upright Man was evolving into a primate with a larger brain, housed in a less rugged skull. Scientists are not sure when Upright Man gave rise to Modern Man, *Homo sapiens* (Wise Man). Some believe there was

an in-between man (Neanderthal Man). Neanderthal Man certainly lived, but he may have been an offshoot from Upright Man.

Neanderthal Man had died out by 35,000 years ago. Men like ourselves had probably evolved in Africa or Asia at least 15,000 years earlier. They may have reached Europe and America by 40,000 years ago, and Australia by 25,000 years ago. Meanwhile Man's primate ancestors had all become extinct.

Evolution has brought about several important differences between *Homo sapiens* and the other primates. One is the shape of the pelvis, the major bone structure to which the legs are hinged. It is much shorter, so that Man can easily walk upright. Other primates, such as the gorilla, have a longer pelvis which is better suited for walking on all fours.

The human hand is very similar in appearance to that of the higher apes, but it is much

Above: *Homo erectus,* or Upright Man, is thought by many scientists to have been the first true human. Creatures of this type could make stone tools, and remains of their fires have been discovered.

more sensitive, and Man has more control over it. A man's thumb is longer in proportion to his fingers than the thumb of, say, a chimpanzee. Man's greater manual skill is one of the reasons why he has been able to make such progress in his development.

However, the most important difference between Man and the other primates is in the size of his brain. The surface area of the human brain is very much larger than that of an ape. It fits into the skull because it has a great many convolutions or folds. This surface part of the brain is called the cerebral cortex. The cortex is the part of the brain which deals with messages to and from other parts of the body, such as the eyes, and also memory and thought.

Left: *Homo sapiens,* or Wise Man, is Man as we know him today. The species evolved from *Homo erectus,* but scientists have not yet found out when this stage of evolution took place.

The Piltdown forgery

In 1911 and 1912 people searching an English gravel-pit near Piltdown in Sussex found a fossil human skull near an ape-like jaw. Scientists claimed that both had belonged to a 500,000-year-old missing link between Man and his prehistoric ape ancestors.

Scientific tests begun in 1949 proved that neither the jaw nor skull of 'Piltdown Man' was very old, and that the jaw was that of an orang-utan. Someone had 'planted' the bones to fool the scientists.

It may have been the country lawyer who found the site, but a likelier culprit was an Oxford scientist wishing to fool a rival scientist who was a friend of the lawyer.

MAN AND NATURE

The world around us is not so natural and unchanging as it sometimes seems. For thousands of years Man has been making alterations to his surroundings. Most of them have been caused by his ever-increasing need for land—for settlements, for farming and, most recently, for industry.

The result of this long history of man-made changes is that over vast areas of the Earth there are few natural landscapes left. In Europe in particular the countryside is almost entirely a man-made creation. Only in Antarctica and the wilder parts of Asia, Africa, Australia and the Americas is the scene as it was before the first men existed.

All animals have some effect on their environment, but on the whole the changes are too small to be detected. The results of some creatures' presence are more easily seen—the artificial lakes created by beavers' dams, for example, or the close-nibbled grass that is left by thousands of grazing animals. Even these pale into insignificance beside Man's effect on the landscape.

Sometimes Man has worked with Nature to produce a new beauty—the green and gold patchwork of an English farmland scene, for instance—but often Man's attempts at progress have been for the worse. His cities bury the land under a mass of brick and concrete. The refuse and effluents from towns and factories poison the ground, the sea and even the air.

In addition to altering the world in which he lives, Man has also interfered with the Earth's plants and animals. Many species have become extinct as a result of his activities, and others have been artificially adapted to suit his needs.

This chapter examines some of the ways in which Man has changed the world he lives in, and the efforts he is now making to try to ensure that any bad effects of those changes are reduced.

A striking example of the way in which Man has altered the face of the Earth – the elaborate pattern of a major highway intersection. Even before this convoluted roadway system was built, the landscape was entirely Man-made, with fields and grass replacing the forest which once clothed this part of the world.

Man and His Environment

Until about 10,000 years ago Man made very little impact on his environment. He was a traveller, wandering from place to place in search of food. Then, about 9000 BC, some people ceased their nomadic life and began to found settlements. They grew crops, and established more permanent places to live. To make space for his farming activities Man had to clear land of its natural vegetation. The simplest method, known as slash and burn, is still practised by primitive peoples in parts of Africa and South America. They cut down the trees and burn the debris. The ash helps to enrich the soil for their crops. This was the method used by the pioneer families of North America from the time of the first settlements.

In its simplest form the slash and burn method of farming is only a temporary one. After a few years the soil is worked out, and the people move on to a fresh area of forest. Some sites are naturally rich, and farmers were able to work the soil over long periods in areas which had their soil renewed annually, such as the Nile valley. The Nile used to flood every year, depositing fresh soil on the land either side. Man's activities have greatly reduced that flooding during recent years.

Right: Constructing the Rajasthan Canal in north-western India. The canal carries water through sandy desert areas to make more land available for farming.

Below: A landscape in process of change – a land reform project in Valle del Cauca department, Colombia. Already the natural landscape of trees is being cleared ready for planting with crops.

years. For though there were people living in all these areas in earlier times, their ways of life were so simple and their numbers so small that they made little impact on the land.

It was not only the clearance of the land that changed its appearance. From very early on people began to establish boundaries between the territories to which different tribes laid claim, and between one field and another to keep out animals. Many of these boundaries have lasted for a very long time. Part of the patchwork of fields that makes up so much of the English landscape, for example, was established in prehistoric times.

Because Mankind is a quarrelsome species, boundaries were also marked out for defensive purposes. Western Europe is dotted with the remains of giant ditches and earth banks, some forming complete forts, particularly on hill sites. Cities were surrounded by defensive walls, many of them built of stone. Stone has to be quarried, and so we see further relics of Man's activities. The scars left by quarrying are in some places thousands of years old.

Quarrying for stone marks the beginning of industry. Soon people began digging in the ground for metals they could fashion into domestic articles and weapons—gold, silver, copper, tin and iron.

Farming, dwelling in cities, industry—all these ways of life depended on a good supply of water. The ancient Egyptians, together with their contemporaries all over the Middle East, dug canals to lead water from the rivers to irrigate their fields. Power from the first water mills was harnessed as long ago as 3000 BC. A good and constant head of water is needed for a water mill, so—like the beavers —people built dams to make artificial lakes and mill-ponds. That was the start of the process by which, today, countless rivers are dammed to ensure water supplies for drinking, industrial processes and power.

Transport, too, has played its part in changing the landscape. For a long time water—rivers and the coastal seas—provided the easiest and indeed the main travel routes. Soon canals were used not just for irrigation but to link rivers and provide short cuts where there were no natural waterways. The biggest era of canal-building has been the past 250 years, and it is still going on, resulting in major waterways such as the Suez, Panama and Kiel canals, and the St Lawrence Seaway.

Ships need harbours. There are many natural ones, such as the Grand Harbour in Malta, but over the years Man has improved them and created artificial ports.

Roads and railways have increasingly carved up the countryside, altering the contours of the landscape. Many roads follow routes that were pioneered by Stone Age men.

As the world's population slowly grew, so more and more land was brought under cultivation. Progress varied from one part of the world to another. The first large-scale settlements, leading to the building of cities, were in Asia and North Africa. Europe developed next, from the period of the Roman Empire onwards. North America, southern Africa, Australia and most of Latin America have been settled within the past few hundred

Above: A completely man-made landscape: the English coast at Gosport, Hampshire has been transformed to provide adequate harbour facilities and the industrial back-up the harbour needs.

177

Pollution

Pollution, the contamination of air, land and water by all kinds of chemicals—poisonous gases, waste material and insecticides, for example—is one of the greatest problems facing the world today. It has upset the balance of nature, destroyed many forms of wildlife and caused a variety of illnesses. Although pollution is greatest in industrial countries, it occurs in every country on Earth.

Most people are only too familiar with polluted air. In heavily industrial areas, fumes from car exhausts and thick smoke from factory chimneys pour forth in clouds, darkening the atmosphere, reducing visibility and making the air unpleasant to breathe. Large-scale burning of fossil fuels—such as coal, gas and oil—in homes and industry produces a wide range of pollutants. These include sulphur dioxide, which damages plants, destroys buildings and affects health; carbon monoxide; nitrogen dioxide; and dirt particles.

Normally the fumes disperse in the air, but sometimes they are trapped by air layers of different temperatures. The result is a fog-like haze known as smog. Britain and other countries introduced smokeless zones and smokeless fuels some years ago and smog no longer occurs, but it still remains a very dangerous problem in Japan and the United States.

The motor-car is a major source of pollution. In Los Angeles, where some 5,000,000 cars pass in and out of the city daily, the level of carbon monoxide in the air is dangerously high. On calm days in the city the fumes settle near ground level. Fumes from car exhausts also pour out lead and nitrogen oxide.

The testing of nuclear weapons, and the peaceful use of atomic energy, have exposed some people to levels of radiation that are too high for safety. Crop spraying by aircraft adds chemical poisons to the air.

Domestic rubbish is another very serious pollution problem. The average American citizen throws away nearly one tonne of rubbish each year. Much of this consists of plastic, metal and glass packaging that cannot be broken down naturally. Instead it lies, with old refrigerators, broken washing machines and abandoned cars, in huge piles for years without decaying. Each year the problem of rubbish disposal becomes more serious.

Sewage causes another form of pollution. Most of it flows straight into rivers, where it is broken down by tiny bacteria. The bacteria need oxygen for this process, but because of the vast quantities of sewage, the bacteria use up the available oxygen in the water, causing the death of countless fish and other river life. Rivers also provide a very convenient outlet

for industrial waste, as well as a source of cooling water for nuclear and other power plants. The result in waters such as the Great Lakes of North America, has been severe pollution which has drastically reduced the fish and plant population.

Like rivers, the oceans have been used as vast dumping grounds for waste of all kinds. One of the most recent sources of pollution is oil: millions of tonnes of it spill into the sea each year. Since the 1960s there has been a spate of accidents involving supertankers. One of the worst was in 1978 when the tanker *Amoco Cadiz* was wrecked on the coast of France, spilling oil over a great area of Brittany beaches. The world's biggest oil spill came in the summer of 1979, when a Mexican off-shore oil well went out of control, sending a huge quantity of oil into the sea and threatening the coast of Texas.

Oil not only pollutes beaches, but also kills fishes and seabirds. Sometimes even more dangerous substances are dumped into the sea. During the 1950s a large amount of industrial waste containing mercury was

Above: Dead trees, killed as a result of air pollution from the giant South Wales steelworks at Port Talbot, seen in the background.

Above right: A young volunteer cleaning oil from the beaches of Brittany, polluted as a result of the *Amoco Cadiz* tanker disaster in 1978.

Right: Once this was a living river. Now it is heavily polluted with sewage effluent, and no plants or animals can live in its foul waters.

dumped into the sea at Minamata Bay, Japan. Marine organisms converted it into methyl mercury, a highly-poisonous substance that damages the central nervous system. Fish and shellfish absorbed the poison, and were themselves caught and eaten. As a result, more than 100 people were poisoned.

Chemicals used for killing insect pests on crops and in the soil are causing serious concern. Among them is DDT (which stands for dichloro-diphenyl-trichloroethane). If taken in sufficient quantities DDT poisons animals and even Man. These pesticides are washed into the ground by rain, and eventually find their way into rivers and lakes. There they are absorbed by small organisms which are eaten by fish. As larger and larger animals in the food chain eat creatures that have absorbed the poison, they accumulate larger doses of the poison. DDT has led to the decline of many predatory birds, such as hawks and falcons. More than 50 species are now in danger of extinction, including the peregrine falcon and the bald eagle—the symbol of the United States.

Conservation

The idea of designating land for the protection of nature and its wildlife is not new. In medieval Europe huge areas of forest land were set aside where the killing of animals by unauthorised persons was forbidden by law, and this type of practice may well go back much further to ancient civilisations. These early measures were concerned only with conserving animals for sport.

The modern idea of conservation, that plants and animals shoud be preserved for their own sake, is little more than 100 years old. It began in North America where, during the 1800s, once-numerous species such as the American bison, the elk, the grizzly bear and the cougar disappeared at an alarming rate, largely due to over-hunting.

In 1872 the United States government set up the Yellowstone National Park, the first of its kind in the world. This set the pattern for a new type of conservation in which animals were preserved, together with their natural habitats. Various national and international organisations to further this ideal were set up. Today the main international organisations include the International Union for Conservation of Nature and Natural Resources (IUCN) founded in 1948, and the World Wildlife Fund, founded in 1961.

Since 1872 thousands of parks and reserves of all sizes have been established by governments throughout the world. They include game reserves, forest reserves, safari parks, areas of outstanding natural beauty and marine reserves. Some are massive. The world's largest national park is in Greenland and covers 8,000 square kilometres (2,000,000 acres). Others are much smaller and exist to protect just a few individual species. In China, the Hsifan Reserve was created as a home for the giant panda, and in Indonesia reserves have been created to preserve the world's largest flower, *Rafflesia*.

Animals cannot simply be kept in protected areas and left alone. There is a constant danger that some species, guarded from attack by their natural enemies, will over-breed and use up available food resources. This has happened with elephants and hippopotamuses in East Africa, and deer in Scotland. In these cases sensible thinning out is essential for a species to thrive.

Zoos now play a part in conservation. Once they were merely places of entertainment where animals were kept in unhealthy, cramped conditions. Today they are increasingly concerned with research and education. In the biggest zoos animals are kept in as near natural conditions as possible and in this way much is learned about their behaviour and requirements. Zoos have also helped to preserve animal life by breeding rare species and returning them to the wild. There have been many remarkable successes, notably the ne-ne or Hawaian goose. In 1950 fewer than 20 existed in the wild. Some were sent to Britain where they were bred in the Wildfowl Trust's sanctuary at Slimbridge. They were then re-introduced to Hawaii.

In much the same way the European bison was saved from complete extinction in the Bialowieza Forest in Poland. Another, more recent success, has been the Arabian oryx. By 1960 this beautiful animal had been almost completely wiped out by hunters. A few were saved and taken to Phoenix Zoo, Arizona. They began to breed and today two small herds exist in zoos in the United States. There are also a few still living wild in Arabia.

Despite these successes some 400 species of

Right: The saiga antelope of Kazakhstan, in the Soviet Union, was saved by a Russian government law banning hunting of it in 1919, when only a few were left. Now there are more than 2,000,000.

Below: This sleepy giraffe has been doped by a drugged dart and is being roped. Like other wild animals in Kenya, it is being moved from a danger area to a safer home in a game reserve. Hundreds of endangered animals have been saved in Africa by this means.

Above: Père David's deer, a Chinese species, was almost extinct in 1900. It was saved by the 11th Duke of Bedford, who bought all the surviving animals from zoos and made one herd of them at his English estate. There are now more than 600, and a herd has been started again in China.

animals are still threatened with extinction. Most of them, notably whales, seals and crocodiles, have been put at risk by the commercial trade in their skins and by-products. Gradually international legislation is being introduced to curb the hunting.

Hunting is not the only cause of wildlife destruction. Much more devastating is the complete destruction of natural habitats by intensive agriculture, urbanisation, industry and technology. Today people everywhere are becoming increasingly aware that conservation covers the whole environment.

Plants in peril

About 20,000 species of plants are threatened with extinction. The cultivation of land, the building of roads and railways and the growth of towns have all taken their toll of plant life.

On a smaller scale some wild flowers are in danger of disappearing because people have been too careless about picking them. If all the flowers are picked from a plant it cannot make seeds to form new plants, and the species dies.

In cultivated areas it is usually necessary to destroy wild plants —the so-called weeds—but wild flowers in their natural environments should never be picked. They are not only beautiful, but they play a vital rôle in maintaining the balance of nature.

Growing Crops

People have been eating plants as a basic part of their diet since the very earliest days of Man's existence, but they only started cultivating them about 10,000 years ago. The discovery that plants could be cultivated was probably an accident. People gathered wild plants to eat; new plants sprouted in the places where seeds from these plants fell on the ground. By observing this natural process people learned to plant seeds deliberately and harvest the resulting crops.

Farming began in the fertile regions of the Middle East—around the Tigris and Euphrates rivers in Iraq, and in Palestine. It also developed independently in parts of China, India and Central and South America. Cereals such as wheat and barley in the Middle East, rice in Asia, and maize (corn) in Central and South America were the first and most important crops. Sorghum was one of the earliest cereal crops to be cultivated in tropical Africa. Wherever crops were sown the first permanent human settlements began to develop. These in turn became the basis for future civilisations.

Plants raised for food included peas, lentils, millet, soya beans, peppers, squashes, root crops and fruits. Most of the important food crops grown now were first domesticated in those ancient days, but many of them have changed considerably. Wheat has altered much through cultivation. The head of wild wheat has brittle spikes which break easily, releasing the ripe grains before they can be harvested, and these grains are enclosed in hulls which are difficult to thresh. Modern wheat species have flexible spikes and loosely-hulled grains.

Some cultivated plants have changed so radically that we do not know what their wild ancestors looked like. Maize, like all cereals, is a member of the grass family, but it has been crossed so many times that today it has no close relative among the wild grasses. It has been so intensely cultivated to suit Man's needs that it cannot disperse its own seeds.

Even in ancient times farmers were concerned with improving the quality and yield of their crops. Seeds for planting were selected from the best, healthiest and most popular plants. By crossing one plant with another a hybrid species was produced that had certain desired characteristics. This sort of selective breeding and cross-cultivation not only produced stronger, more prolific plants, but also resulted in a wide range of food crops. For example, people eat many different varieties of apples, but all of them are descended from one species—the wild crab-apple. The cab-

Left: The huge bunches of large grapes which this Cypriot girl is picking are the result of years of plant breeding.

Right: Two Japanese women pause for a moment during their work in the rice paddy fields, where they toil knee-deep in mud. Rice is the staple food of Asian countries.

Below: Modern farming in North America where machines do much of the work and save time.

Below right: Raising water for irrigation by means of an Archimedean screw – an ancient device still used in Egypt.

bage family, too, has many members, including red and green cabbages, turnips, Brussels sprouts, kale and cauliflower. All were developed from one kind of wild cabbage.

Yields from early crops were tiny compared with those achieved today. In about AD 1200 the harvest from one hectare (2.5 acres) of wheat was enough to feed four or five people for a year. These days a well-tended hectare in good wheat-growing country might yield enough to feed 40-50 people for one year.

During the 20th century crop cultivation has become a science, and massive increases in yields have come about through scientific research. Natural and artificial fertilisers enrich the soil; sprays protect crops from pests and diseases, although it is now known that pesticides do have dangerous side effects (see pages 178-179); and chemical growth-hormones enable farmers to control the development of crops. They can be used to boost or retard growth. Wheat prevented from reaching its natural height by the use of growth-hormones may resist storm damage and be easier to harvest.

An important achievement has been the development of fast-maturing strains of plants —the so-called 'green revolution'. By cross-breeding, dwarf varieties of wheat, maize and rice have been bred which give very high yields, and in some cases produce two or more crops in a single season. The new strains of rice have meant record harvests. Traditionally in India there is only one rice crop a year. With the new types of grains two, three or even four crops a year are possible.

What is a weed?

Weeds are plants that grow where they are not wanted. In cultivated areas, in farms and gardens, for example, they can be a constant nuisance, using up space, sunlight and nutrients needed for crops.

Yet when they are growing in the wild, what a farmer would call a weed is not only beautiful but also a vital part of the ecological chain, providing food for insects and other animals and enriching the soil.

The idea of what constitutes a weed also differs around the world. In Europe and North America burdock is considered a weed; in Japan it is eaten as a vegetable. When wheat cultivation began some 8,000 years ago oats and rye, which today are major cereal crops, were regarded as weeds.

Rearing Animals

People have always relied on animals—for food, clothing, labour, transport and even companionship. This dependence on domestic animals helped to shape Man's development.

The first attempts at capturing and taming wild animals probably occurred about 15,000 years ago while Man was still a nomadic hunter and food-gatherer. Probably the dog was domesticated first. It was used for hunting and guarding, and its flesh was eaten. Then goats and sheep were reared for their milk, skin and wool. They were kept in small flocks that could be herded from place to place. The earliest known remains of domestic sheep date from 8500 BC.

As agriculture became established, Man began to keep cattle, pigs, donkeys and horses. The llama and cat were then domesticated, and about 3000 BC the camel was brought under Man's domination. The ancient Egyptians kept bees for their honey, which was the chief food sweetener until about 500 years ago, when sugar became more generally available. Silkworms were bred in China by about 3000 BC and in what is now Pakistan chickens were domesticated in about 2000 BC, ducks and geese a little later.

Most modern domestic animals look very different from the wild creatures that were first tamed. As with plants, this is due entirely to the intervention of Man. By cross-breeding carefully chosen animals, Man has been able to produce many domestic varieties from just one wild species. Many of these have qualities or characteristics that are not found in the wild but that have been deliberately introduced. The dogs, horses, cattle, sheep and poultry of today are all the result of a long process of selective breeding.

Modern dogs are probably descended from the wild grey wolf of Europe, Asia and North America. Wolves may have been attracted to early human settlements by scraps of food lying around. Being social animals, they gradually became Man's first close animal companions. As they were domesticated, physical changes occurred. They became smaller, their tails began to curve upwards and their coats began to vary in colour. Their teeth also grew smaller and a primitive type of dog began to emerge.

At some time Man began to breed dogs for specific purposes, as hunting dogs, guard dogs, work dogs and pets. Today there are more than 100 varieties of all shapes and sizes. Selective breeding still continues, particularly for show purposes. The specific characteristics that breeders aim for are not necessarily ones that would help an animal to survive in the wild, and many of the pedigree dog breeds suffer from diseases and weaknesses caused by the search for prize-winning or fashionable features.

The horse has probably been the most valuable of all domestic animals for transport and for farm work. The wild ancestor of the horse was the tarpan, a small animal that roamed wild through central Asia. Early domestic horses were not much larger than donkeys, but today the breeds range from heavy Shire horses to sleek racehorses and tiny ponies. Only one wild species still exists—Przewalski's horse which lives in Mongolia.

Most modern European and American cattle are descended from the aurochs or wild ox, a formidable animal that stood some 2 metres (6½ feet) high. The result of hundreds of years of careful selection has resulted in a wide range of highly specialised breeds. The best known include the French Charolais and the Aberdeen Angus, both bred for beef, and the Jersey cow, famous for the rich quality of the milk it produces. Early wild cows probably produced less than a litre a day; now an average European cow can give about 25 litres (5½ gallons).

The wool yield of modern sheep is far heavier than that of primitive domestic sheep. Chickens are all descended from the wild Indian red jungle fowl, and were originally bred for sport and meat. During the past 30 years the main concern has been to produce intensive egg-laying breeds, and the rearing of poultry has become a major industry.

Animals of the future

Animals that may be of value to Man in the future include the beefalo, an experimental cross between the American bison and domestic beef cattle; the African eland, which thrives on poor land and can be domesticated for its meat; and the sea lion, whose diving ability has already been put to use by the United States Navy for carrying messages down to human divers.

Glossary

Abdomen Region of vertebrate body between chest and pelvis; the hind part of an arthropod.

Aestivation The summer sleep of some animals.

Amino-acids Fundamental ORGANIC COMPOUNDS making up the PROTEIN MOLECULE.

Anal fin Fin at the lower rear of a fish.

Antennae The pair of feelers on the head of an arthropod.

Anther The part of the STAMEN producing POLLEN grains.

Antheridium Male sex organ of algae and fungi.

Archegonium Female sex organ of ferns, liverworts, mosses and most gymnosperms.

Atom The smallest possible unit of an ELEMENT.

Baleen Whalebone; gristly plates inside the mouth of some whales, used to filter food from the sea.

Biped Any two-legged animal.

Camouflage Shape or coloration of animals which makes them hard to see against their background.

Carbon dioxide A gas given off by animals when they breathe out.

Carnivore A meat-eating ORGANISM.

Cartilage Gristle; flexible part of skeleton, as in nose and ears.

Caterpillar LARVA of a moth or butterfly.

Cell The basic unit that makes up animals and plants. Growth and reproduction result from cell division.

Cellulose Fibrous material forming the CELL wall of plants.

Chlorophyll The green PIGMENT in plants, needed for PHOTOSYNTHESIS.

Chloroplast Photosynthetic part of a CELL, containing CHLOROPHYLL.

Chromatin The part of the CELL NUCLEUS that forms the CHROMOSOMES.

Chromosome Thread-like structures in the CELL NUCLEUS, containing GENES.

Chrysalis PUPA: the resting phase between the LARVA and the adult of some insects.

Cilium Fine hair-like projection from a CELL surface.

Cocoon Envelope spun by some insects to guard their EGGS or PUPAE.

Commensalism A close sharing between members of different species.

Compound A substance made up of two or more ELEMENTS, such as water.

Coniferous Bearing cones and evergreen leaves.

Cuticle The non-living outer layer of an animal or plant.

Cytoplasm The PROTOPLASM surrounding the NUCLEUS of a CELL.

Deciduous Shedding leaves each year at the end of the growing season.

Dicotyledon A plant with two seed leaves in its EMBRYO.

Dimorphism Existence of two distinct forms in the same species.

Dorsal fin The fin on the back of a fish.

Egg The seed produced by a female animal to be fertilised by a SPERM.

Element A substance made up of only one kind of ATOM, such as OXYGEN.

Embryo An immature animal before it is born or hatched; a young plant contained in its seed.

Enzyme One of many PROTEINS produced in CELLS that set off chemical changes in the body.

Epiphyte A plant that grows on another plant for support, but not as a PARASITE.

Exoskeleton A protecting support on the outside of an animal's body.

Fermentation Slow breaking down of an ORGANIC substance by ENZYMES or ORGANISMS such as yeast.

Flagellum A whip-like projection from the surface of a CELL. The motion of flagella enables tiny ORGANISMS to move around.

Fossil Remains or cast of former ORGANISM preserved in rock.

Frond The leaf of a fern or palm; the leaf-like part of a seaweed.

Gene A unit of heredity carried on a CHROMOSOME.

Germination The beginning of development of a plant EMBRYO.

Gills The breathing ORGANS of animals which take in OXYGEN from the water.

Gland An ORGAN that produces a certain substance in the body, such as HORMONES, SALIVA or tears.

Habitat The normal area lived in by a plant or animal.

Herbivore A plant-eating creature.

Hibernation The winter sleep of some animals.

Hormone A secretion from a GLAND, controlling a bodily function.

Hybrid An ORGANISM which has parents whose GENES are not alike.

Hypha One of the fine threads making up the body of a fungus.

Imago An adult insect which has passed through METAMORPHOSIS.

Inflorescence The part of a plant which carries the flowers.

Inorganic Not produced by or having the characteristics of ORGANISMS.

Instar The stage of insect development between two successive moults.

Invertebrate An animal without a backbone.

Larva The immature, active state of development of some animals that pass through METAMORPHOSIS.

Lungs The breathing ORGANS of animals which take in OXYGEN from air.

Mandible The biting mouth-part of some INVERTEBRATES; the lower jaw of VERTEBRATES.

Mantle In animals, a fold of skin that secretes a shell.

Marsupial A mammal whose immature young grow in an external pouch.

Meiosis CELL division in which each resulting NUCLEUS has half the number of CHROMOSOMES of the parent cell.

Membrane A thin layer of tissue connecting CELLS and ORGANS.

Metamorphosis The series of changes that takes place during the growth of some creatures, especially insects.

Mineral Any solid INORGANIC material of a particular chemical composition.

Mitosis CELL division in which each resulting NUCLEUS has the same number of CHROMOSOMES as the parent cell.

Molecule The simplest unit possible of any chemical COMPOUND, made up of ATOMS of ELEMENTS.

Monocotyledon A plant with a single leaf in its EMBRYO.

Mould Fungus that causes mildew —the growth on stale food.

Moult To shed feathers, hair or outgrown CUTICLE.

Mucus Slimy protective coating, secreted by GLANDS.

Mutualism A close association between members of different species, in which both partners benefit.

Nectar Sweet fluid produced by flowers and attractive to insects.

New World North and South America.

Notochord A supporting rod taking the place of a backbone in EMBRYOS and some INVERTEBRATES.

Nucleus The chief ORGANELLE of a CELL, containing its CHROMOSOMES.

Nymph An immature stage of some insects. It develops into an adult without becoming a PUPA.

Old World Europe, Africa and Asia, the parts of the world known before the Americas were discovered.

Omnivore A creature that eats both plants and animals.

Organ Any part of an ORGANISM that has its own job to do, such as a heart or a leaf.

Organelle Any part of a CELL that has its own job to do, such as a CHLOROPLAST.

Organic Produced by or having the characteristics of ORGANISMS.

Organism Any living plant or animal.

Ovary Female ORGAN producing EGGS or egg cells.

Oxygen A gas that is an essential ELEMENT of the air we breathe.

Parasite Any ORGANISM which feeds on another organism without giving any benefits in return.

Parthenogenesis Development of an egg without fertilisation by a SPERM.

Peat A mass of partly rotted plant fibres found in boggy ground.

Pectoral fins The pair just behind the head, attached to a fish's 'chest'.

Pelvic fins The pair attached to the 'hips' of a fish.

Pheromone A chemical released by animals which acts as a 'messenger' to others of the same species.

Photosynthesis The production of carbohydrates from CARBON DIOXIDE and water by green plants and some bacteria, using energy from sunlight.

Pigment A substance which gives colour to plant and animal tissues.

Placenta An ORGAN that supplies food for an EMBRYO in the womb.

Plankton Tiny floating ORGANISMS of lakes and sea, eaten by fish and whales.

Pollen The male fertilising powder produced by plants.

Pollination Transfering POLLEN from ANTHER to STIGMA.

Predator Any animal that hunts and kills another animal for food.

Prehensile Adapted for holding on.

Proboscis A prehensile trunk; long flexible mouthpart of insects, used for piercing and sucking.

Protein Complex ORGANIC COMPOUNDS made up of AMINO-ACIDS, essential to all ORGANISMS.

Protoplasm The living substance within a CELL.

Pseudopodium False foot put out by tiny organisms to grasp food.

Pupa An insect at the inactive stage between LARVA and adult.

Quadruped Any four-legged animal.

Radicle Rootlet of a sprouting seed.

Respiration Breathing: taking in OXYGEN and giving off CARBON DIOXIDE.

Sac A bag-like part of an ORGANISM.

Saliva Spittle—the fluid secreted in the mouth.

Savannah Tropical African grassland.

Scavenger A CARNIVORE that finds its food without hunting or killing for itself.

Segment One of the repeated body sections of animals such as a worm.

Sepal A leaf-like part outside the petal of a flower.

Spawn A cluster of EGGS laid by fish and amphibians; the reproductive part of fungi.

Spectrum The band of all the colours which make up white light.

Sperm Male reproductive CELL, produced to fertilise an EGG.

Spore Reproductive body of lower ORGANISMS such as fungi and bacteria.

Stamen Male reproductive ORGAN of a plant, carrying the ANTHER.

Stigma The female part of a flower where POLLEN lands and starts to grow.

Stimulus Any action or change that makes a plant or animal react.

Stratum A layer of CELLS or rock.

Succulent A juicy plant that stores water in its tissues, such as cactus.

Symbiosis A close association between two different types of ORGANISM.

Taiga The region of coniferous forest in the marshy land of North America and northern Siberia.

Taxonomy The naming and classification of plants and animals according to their similarity of structure.

Temperate Having a mild climate.

Tentacle A slender whip-like organ of plants and INVERTEBRATES, used for moving, feeding or grasping.

Thorax The chest: the part of the body between the head and ABDOMEN.

Thyroid A gland in the neck of vertebrates that secretes HORMONES controlling body growth.

Tissue A layer or group of CELLS that together form special functions.

Transpiration The loss of water vapour from a plant, mainly through its leaves.

Tundra The treeless zone lying south of the Arctic ice cap.

Valve One of the pieces forming the shell of a bivalve shellfish.

Vertebrate Animal with a backbone.

Whorl An arrangement of leaves or petals around the stem of a plant.

Index

Compiled by Richard Raper, B.Sc.

Page numbers in **heavy type** indicate a drawing or photograph

Picture Credits

Heather Angel; Ardea; Aquila; Barnaby's Picture Library;
John Bulmer; Cyprus Ministry of Information; FAO Rome;
Fox Photos; Frank Lane; Marion Morrison; Oxford
Scientific Films; Picturepoint; President and Council of
the Royal College of Surgeons of England; Sealand Aerial
Photography; Science Museum, London; Syndication
International; John Topham Picture Library; ZEFA;
Zoological Society of London.